Character Design
for Mobile Dev

RotoVision

A RotoVision Book

Published and distributed by RotoVision SA
Route Suisse 9
CH-1295 Mies
Switzerland

RotoVision SA
Sales and Editorial Office
Sheridan House, 114 Western Road
Hove BN3 1DD, UK

Tel: +44 (0)1273 72 72 68
Fax: +44 (0)1273 72 72 69
www.rotovision.com

10 9 8 7 6 5 4 3 2 1

ISBN: 2-940361-12-6

Art Director Tony Seddon
Design by Studio Output

Reprographics in Singapore by ProVision Pte.
Tel: +65 6334 7720
Fax: +65 6334 7721

Printed in Singapore by Star Standard Industries
(Pte.) Ltd

Character Design for Mobile Devices

Mobile games, sprites, and pixel art

NFGMan

Contents

006
1: Introduction

014
2: History of portable devices

044
3: Theory—changing hardware

078
4: Mobile developers

132
5: Genre / pixel histories

Introduction

Welcome to the first book on the designs, the technologies, the issues, and the techniques behind character design for mobile devices and games.

Video games are no longer a niche phenomenon. It's been a decade since the Sony Playstation was unleashed upon the world and videogames secured their place at the table of popular culture. They're here to stay; no one disputes this now. Nearly everyone has a video-game system in their house, either in the shape of a computer or a dedicated game console, but there's another new wave. A portable wave.

It's hard to go anywhere now without seeing someone playing on a portable device—their phone, Personal Digital Assistant (PDA), or dedicated portable game system. It started with Nintendo's GameBoy, and nearly everyone has had a go at the portable space. Sega, NEC, Bandai, SNK, and now

Sony have all tried their hand with portable systems. Many unknowns have given it a shot too: GamePark, Gizmondo, and Tapwave to name but three. Even Microsoft and Palm have ensured their PDAs are game-friendly.

Portable gaming is available to everyone; the hardware is accessible and very affordable. There is a long history of portable gaming and its legacy is impressive. The graphics are impressing anew with every hardware generation.

This book is a tribute to 2D gaming graphics and portable game graphics.

Please enjoy.

1. Blue Slime, Chris Hildenbrand

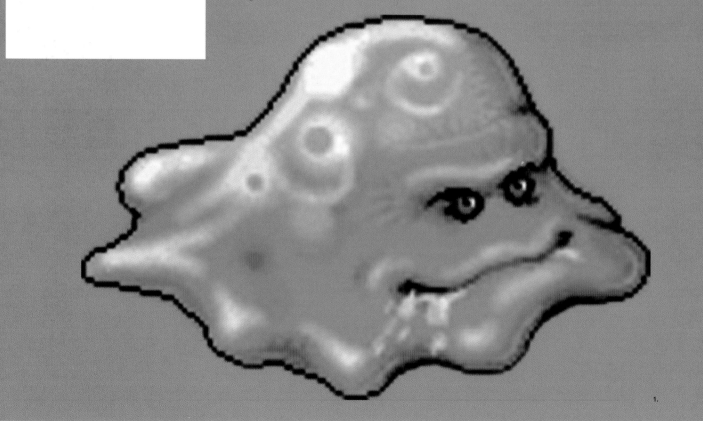

1.

Introduction

This is a nameless blue slime, drawn by Chris Hildenbrand, for a role playing game (RPG) that was never released. Slimes are an RPG staple, but they're rarely drawn with such character.

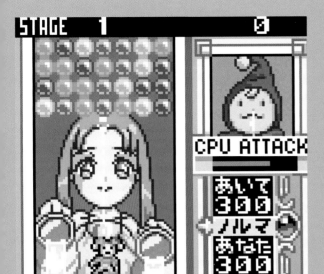

1. *Magical Drop*
2. *Super Mario World*
3. *Rainbow Island*

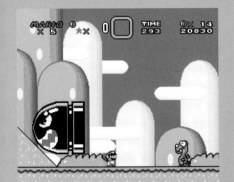

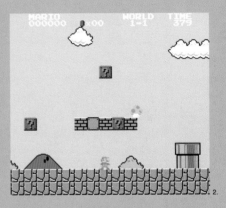

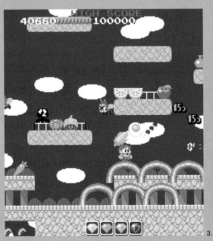

Introduction

What is a pixel?

Pixel: //n.//
An abbreviation for pixel element, the smallest element of a computer image.

With very few exceptions computers have always generated their images one pixel at a time. For sure modern hardware does it faster, smoother, and with more pixels than ever before, but the concept remains unchanged.

Images are made out of these primary components one square or rectangle at a time inside the computer, and they're displayed this way no matter if you're playing them on a TV, monitor, plasma, or LCD screen. Each pixel represents the smallest component of a computer image and is made of the primary additive colors red, green, and blue.

The pixels are tightly packed and often you don't notice them. In fact, modern games and hardware go to great lengths to hide the pixels, with new-fangled technologies like anti-aliasing. Back in the old days you couldn't hide your pixels. The Atari 2600, grand-daddy of consoles, wouldn't dream of hiding any of its pixels: with fewer than 200 of them on every horizontal line none could be spared. Before the Sony PlayStation the only way to make individual pixels less obvious was to carefully prepare them while developing the game. Making them less apparent on the fly, in real time, was a crazy dream.

All that changed when video games moved to 3D. Game designers no longer needed to place each pixel individually. Left to their own devices, computers tended to run amok with polygons, and the pixels became ugly. Game design shifted

1. The Beatles, Craig Robinson

1.

What is a pixel?

It's not easy creating something recognizable when your entire canvas is a mere 32 pixels tall, but few would be unable to recognize the four Beatles, above. "Minipops" of hundreds of bands are available on Craig Robinson's flipflopflyin.com Web site.

from hand-crafted to mass-produced graphics. Frame rates increased, realism crept in, and polygons took over. The age of pixel artistry had finally ended.

Or had it?

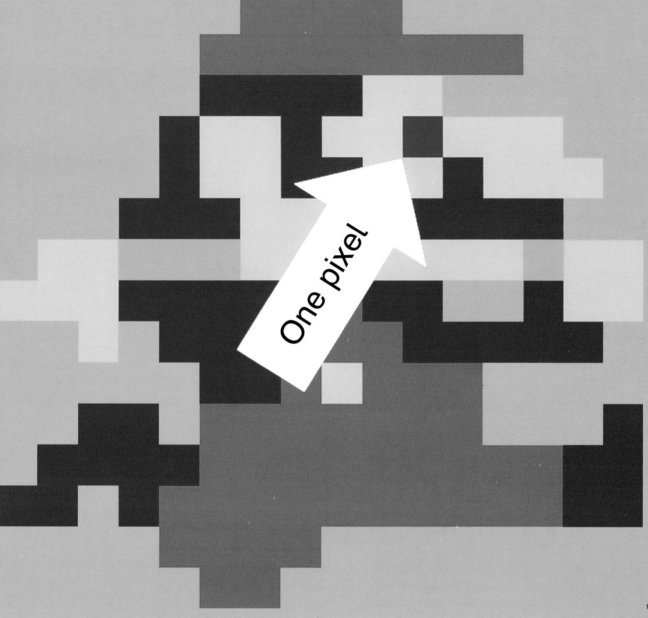

One pixel

1.

What is a pixel?

Pixel art is traditionally created one pixel at a time. It's not easy, and it's not fast, and after working in close up with giant squares it can be difficult to envisage the larger work.

Pixels as art

First the technology limited the designs; then people saw this as an advantage in creating great designs and artworks.

As newer consoles grew in power and capability the emphasis on game graphics shifted. Once a graphic artist might have stayed awake at nights trying to create that perfect character, but modern game designers looked for shortcuts at every opportunity. New images could be drawn or painted and then scanned in, easing the designer's workload and greatly speeding development.

Game companies were always trying for the next new thing, trying to impress gamers, occasionally at the expense of gameplay. LaserDiscs were used in arcade games to present truly amazing graphics, created by traditional Disney-style animation or shot with real movie cameras. The pixel was forgotten for this short time—the game imagery was as good as any movie or television show, but the games weren't much fun. An elusive concept at best, gameplay

might best be described as the "fun factor." These LaserDisc games were far less interactive and enjoyable than their predecessors, and they quickly fell out of favor.

Digitized graphics were the next labor saver. Companies like Williams and Atari found that they could digitize photographs to create images with stunning realism, but again there were limitations. The images produced tended to be ugly and poorly animated.

Then came the polygon games, with miraculous new hardware that could spin and scale and animate 3D objects smoothly and with thousands of colors. Still the designers struggled. Most early games used very few polygons to reduce the load on the processor and increase animation speed and smoothness. The first 3D games tended to resemble a work of geometric abstract art more than

1. *Vampire Savior*, Capcom, arcade

2. *Aurail*, Sega / Westone, Arcade

3. *Metal Slug 2*, SNK, arcade

1.

2.

3.

Introduction

Some people would claim that game character design is not art. And yet it's surely hard to argue that these images were not created by artists.

anything else. Now modern consoles are wickedly powerful indeed, capable of using hundreds of thousands of polygons on every screen—a veritable plethora of realistic 3D worlds and objects for players to manipulate.

Throughout all of these improvements and technological advancements one constant has remained: the pixel artist—the hardworking person who labors endlessly, creating images one dot at a time—has had less and less to do with what's seen on the game screen. Perhaps not unlike the farrier whose horse-shoeing business declined with the advent of the motorcar, the pixel artists' time is drawing to a close. Or is it?

There's a growing demand for pixel artists again, thanks in part to a growing demand for portable games. Creating a game character from a

series of polygons is so totally different from creating one from pixels that they might as well be different media entirely. In fact, except for the final medium—the video screen—they are completely different. There is no crossover between the tools needed for pixel or polygon art, and there are very few artists proficient in both styles.

There's a certain style, a distinct charm, to pixel art. It's very clear and obvious, even when presenting a complicated image. It evokes a certain retrograde nostalgia, memories of games played long ago, and a part of our culture that has almost been forgotten. Fans of retro gaming point to pixel graphics as a primary reason for eschewing modern games. To many people, pixel graphics simply look better.

It's not just the retro fans either; a new generation has started manipulating pixels with a surprising fanaticism. And it's not just for guys. Video games have traditionally been a male-dominated passion, but pixel art appeals to both sexes and all ages. Pixel art is big business too. Some designers regularly appear in the pages and on the covers of magazines.

It might almost seem that pixel art, freed from its legacy as an artifact of old games, has come into its own as a legitimate art form.

1. *Street Fighter Alpha 2,* Capcom, arcade

2. *Street Fighter Zero 2,* Capcom, arcade

3. *Blazing Star,* SNK / Yumekobo, arcade

4. *Samurai Shodown 3,* SNK, arcade

1.

2.

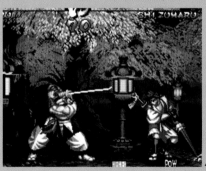
3.

4.

Pixels as art

Game graphics cover a massive range of styles, from abstract or minimalist imagery to fully realized recreations of real locations, to realistic-looking places that never existed.

Pixels compared to vectors

As with any pixel-based design work, size rapidly becomes a problem, whereas vector-based work does not have these drawbacks. What are the issues?

Pixels don't scale well. A pixel is a hard element that can't be stretched at all. If the screen is ten pixels wider than the game screen, then ten new pixels must be added. This has long been a drawback of pixel art, and is one of the most significant, time-consuming considerations when porting graphics to a different platform.

The final output for any modern game has to be pixels. All LCD, CRT, and plasma screens work in pixels. They can't handle anything else, so portable devices, and in fact all LCD screens, cannot easily stretch an image. They can fake it, but the result is very processor intensive, and the resulting images are blurred and suffer degradations in contrast and saturation.

Pixels, however, aren't the only way to create graphics with a computer. Modern flat screens, both plasmas and LCDs, are

designed to imitate raster scan CRT displays. At the risk of getting technical, a brief description is in order.

In a black and white raster monitor a single beam of electrons shoots from the rear of the monitor, through a powerful electromagnet, toward the front. The magnetic field steers the electron beam in horizontal sweeps, from the top to the bottom of the screen, over and over again, until it reaches the bottom. It then quickly jumps back to the top, and begins again. Each vertical top-to-bottom sweep draws one frame.

When the electrons strike the front of the monitor they excite a very tiny section of a layer of phosphorous, causing it to glow. If the beam were locked at full power the entire frame would be white. Conversely, if the beam were off the frame would be black.

1. *Squid Love*, Nicky Werner (www.segasaturn.de)

2. Mutron Kun, Lawrence Wright (from *Mutant Storm*, UPL, arcade)

1.

2.

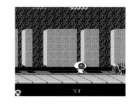

This is Mutron-kun, a character from an obscure 1986 UPL game called Mutant Night. The small sprite cannot be enlarged without making the pixels more and more obvious. Reach a certain point and the image becomes unrecognizable. The eye cannot rearrange it into an image when it has become, to all intents and purposes, a large pile of squares.

By varying the strength of the beam as it sweeps across the screen various shades of gray can be drawn.

A color raster monitor works the same way, except that three beams are now swept across the screen in unison. The front of the screen is coated with three colors of phosphorous, and each beam targets only one color: red, green, or blue. Typically, the timing of these horizontal and vertical sweeps is locked to a small range of frequencies, which limits the number of pixels that can be displayed. If your beam can only cross the screen 240 times per frame, your vertical resolution has a maximum of 240 pixels.

Some of the first video games used the same glass tubes, phosphors, and electron beams, but in a completely different way. These monitors were called vector monitors, and instead of following a horizontal, top-to-bottom scanning pattern they would throw the electron beam around the screen wherever it was needed. This approach allowed for very smooth movement and, because of the vector objects' simplicity, rotation and scaling, which raster games didn't have the power to accomplish. Several companies championed the vector

approach, Atari and Sega being the best known. Games like *Asteroids*, *Star Trek*, and *Gravitar* were big hits.

Vector monitors were fragile and prone to failure—Sega's monitors often caught fire! Vector games therefore disappeared from the arcades within just a few years. The idea of drawing images with lines, however, was a good one. In the mid-eighties computers like the Atari ST and Macintosh utilized line drawings, or vector graphics, to produce a new kind of art.

Pixel, or raster images, don't scale very well. Vector graphics, however, can scale with no limit, without degrading the image quality. Instead of defining every point in an image, a vector image describes the image as points, lines, and colors. These line drawings can be mathematically scaled and deformed in any increment without becoming jagged like pixels will. When the computer displays the vector image it first renders it, calculating each pixel's location before sending it to the display. The end result is a flexible image that takes less space and is more malleable than a raster image. Vector images, like polygons, are very CPU intensive—the larger they are the more pixels they have, and the longer they take to render.

Vector graphics never caught on with video games. Primitive vector arcade games were able to work because the monitor would display each line as it was drawn, but modern monitors draw the vectors on a raster screen, so the entire image must be drawn before it is ready for display. Early hardware had neither the CPU power nor storage space to pre-render the screens this way, so for a very long time vectors simply couldn't be used for games.

Now, however, computers are capable enough, and vector games are once again common—this time as Flash applications run in the common browser. Images can be scaled and deformed in real-time, but again there are tradeoffs: vectors cannot be manipulated with the same fine control that pixels can, so images tend to look bland and generic. Like polygons, vectors require a very different set of skills from the artist, and it seems that for the forseeable future they won't be popular except for niche and/or amateur Web-browser games.

1. NEC PC Engine, Lawrence Wright

2. JVC WonderMega, Lawrence Wright

3. Sharp X68000, Lawrence Wright

1.   2.   3.

Pixels compared to vectors

These images of game consoles were created with lines in a vector drawing package. Vector graphics can be scaled to any size without loss of detail or pixelation.

History of portable devices

Let's scroll back a few decades and see how we got here, and how earlier hardware generations influenced what we can achieve technically today— and, therefore, also in terms of our designs.

Prehistory

Before Nintendo's GameBoy there were many attempts to make games portable. For many reasons, not least the relative crudity of displays and processors, they never caught on. Some attempts were valiant, but most were doomed attempts to push the available technology too far.

At first these games were purely mechanical, such as ball bearings in a maze or an obstacle course that could be tilted or rotated to "win." Tomy produced hundreds of variations on this theme, as did many companies. These quickly gave way to windup and battery-operated toys such as racing games, where a windup spring rotated a track and the player avoided other cars or pitfalls by manipulating a tiny wheel.

One of the first hit portable games using newfangled integrated circuits was Simon, a 1978 release that turned into an eighties phenomenon. It had only four buttons that would light up, and the player had to memorize patterns that changed with every game. Of course it wasn't exciting enough, and every year saw newer and better games and hardware released.

There were two aberrations in the portable scene, the Adventurevision and the Microvision, but neither made any kind of lasting impression. In 1979 the Microvision was released, a strange, handheld machine that used replaceable panels instead of cartridges. In each panel was an entire system— CPU, game data, and buttons— everything but the screen, batteries, and a paddle. It had a very limiting resolution of 16 x 16 pixels, and only 11 games were made. The Adventurevision came out in 1982, a table-top system with interchangable cartridges. It used a remarkable display system, a line of 40 red LEDs, and a spinning

1. Texas Instruments Speak & Math
2. Tiger PlayMaker
3. Tandy Cosmic 3000
4. Milton Bradlay Comp IV
5. Coleco Total Control 4
6. Sears 7-in-1 Sports

Images courtesy of Manu
www.pelikonepeijoonit.net

1.

2.

3.

4.

5.

6.

mirror that gave the illusion of a 150 x 40 display. It was a failure by any standard, and support was dropped after only four games were released. Interestingly, Nintendo resurrected the technology with their Virtual Boy in 1995. This also used mirrors and two single lines of LEDs (one for each eye) to produce 3D images.

Liquid crystal screens at the time were hard to see and slow, so many games tried to use LEDs instead. There weren't any real alternatives, since no other light was battery-friendly and cheap enough. Even though the little red lights could give only the most rudimentary indication of movement, many games such as football and baseball were tried and quickly forgotten.

Some companies tried to make portables with Vacuum Fluorescent Displays, the same kind of technology used in most microwave and VCR displays. These allowed for very bright colors, but few of them—odd shades of red, blue, and green were about all you could expect. They were expensive to manufacture and were made of glass, which made them heavy and fragile. They typically used four or more C- or D-cell batteries, and were "tabletop" games rather than truly portable. Coleco was one of their biggest proponents, with conversions of many arcade hits like *Zaxxon*, *PacMan*, and *Frogger*.

There was not much variety available with this kind of game, but as technology marched on, the games became much more sophisticated. It was Nintendo's Game & Watch series that defined the direction for future portable gaming—cheap and easy on the batteries. These games used LCD screens with black objects that could

be displayed or hidden but couldn't change shape. Fanciful black men avoided vicious black birds or hunted for black gold, guarded by a black octopus. The Game & Watch series was so popular that 59 games were eventually released over an astonishing 21 years. Nintendo held nothing back in the quest to entertain players. Different games with many kinds of gimmicks were tried: multiple screens, transparent playfields, and even miniature arcade cabinets sporting colored LCD displays. Each Game & Watch could play only one game, so if you wanted to play four different games on your vacation you had to bring four different Game & Watch units.

Portable gaming never really took off throughout all of this, and for a few good reasons. While many units were sold—especially Nintendo's Game & Watch series, which eventually sold over 40 million units—all these portables were cheap and essentially throwaway entertainment. You can imagine how Nintendo's engineers, frustrated by the lack of flexibility in the Game & Watch design, made the logical leap to a cartridge-based system with a multipurpose screen. Finally they got it right: Nintendo's GameBoy hit in 1989, and gaming on the go finally achieved the big time.

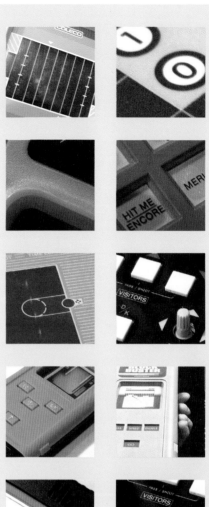

Nintendo Game & Watch

Nintendo's legendary system was one of the prime movers of portable gaming, and today's gaming systems.

Nintendo's Game & Watch series was perhaps the first portable success story. Video games were the trend of the day, and like many toy companies at the time, Nintendo sought to fill a niche. Games of the day weren't portable, and it wasn't a significant leap to imagine a portable game, something that could be played when no arcade or game console was available.

Nintendo made several pioneering moves with the Game & Watch series. They were much thinner than other games of the time and could truly be considered pocket-sized. They were functional too; each one had a tiny clock display and could be used as an alarm. Nintendo had great success with the series, releasing several games based on big licenses such as Snoopy, Mickey Mouse, and Popeye. Several of Nintendo's greatest games had Game & Watch

versions too. *Donkey Kong*, *Donkey Kong Junior*, and *Zelda* all received the treatment. Even *Super Mario Bros*, Nintendo's flagship title, was released as a special-edition prize Game & Watch.

The variety of individual Game & Watch units was limited. The simple LCD screens could display only the objects that were "drawn" on them at the factory. *Donkey Kong Junior*, shown here, didn't allow for a great deal of variety. Junior's entire range of movement consisted of just X positions. When moving he would skip from one position to the next. Each position was carefully drawn to appear static while suggesting a kind of directionless motion. If Junior backed up a step he wouldn't appear to be moving backward. There were no failure poses. If Junior was hit by an enemy he would simply flash.

1. *Oil Panic 2* (Game & Watch)
2. *Donkey Kong 2* (Game & Watch)

1.

2.

History of portable devices

The Game & Watch series covered a lot of ground, and a lot of time. Nintendo was very creative, releasing original games playing to the format's strength, and managing to create surprisingly playable versions of their arcade hits.

The enemies on the screen were the same. Every time they jumped from one place to another the tiny speaker inside the Game & Watch would beep or tick, and players would quickly learn to time their movements by the sound. The game, simple enough at first, became a frantic test of reflexes as the score climbed and the enemies poured onto the screen faster and closer together.

The colored elements were simply printed on a transparent sheet of plastic that sat above the LCD screen, providing a level outline and much-needed color. The games were quite simple but they were challenging and addictive just like classic arcade games were—no variety, but plenty of fun. The Game & Watch series eventually reached nearly 60 titles, and an astonishing forty million units were sold. Its creator, Gunpei Yokoi, went on to create another little Nintendo product called the GameBoy, which was also moderately successful.

1. *Turtle Bridge* (Game & Watch)
2. *Oil Panic* (Game & Watch)

1.

2.

Nintendo
Game & Watch

The Game & Watch enjoyed phenomenal success in dozens of countries. They were Nintendo's first big hit, and for a long time set the standard for portable gaming.

After Nintendo's Virtual Boy failed to sell, Gumpei Yokoi—the man who created the GameBoy and the Virtual Boy—quit Nintendo and was immediately hired by Bandai, the largest toy maker in Japan. Yokoi and Bandai quickly released the WonderSwan, which thanks to Bandai's toy and cartoon licenses quickly saw massive sales in Japan. Despite the release of 196 games and three versions of the hardware Bandai realized that its games sold more units on the GameBoy than its

own hardware, and the WonderSwan was cleared out of Japanese shops in 2004. Perhaps its biggest claim to fame, besides an incredible number of anime-related role playing games (RPGs), was its re-release of Square's massively popular Final Fantasy games.

Nintendo never had a real competitor to its GameBoy. Despite several strong entries from very respectable companies, none could compete with Nintendo's marketing clout or its

unerring ability to determine what gamers wanted to buy. It was seven years after the GameBoy's release that Nintendo released the GB Pocket, a slimmer GameBoy with a much better screen. Nintendo followed this with the GB Color in 1998, which offered, besides a nice color screen, slightly more CPU power. Nintendo finished the design for the Project Atlantis, a GameBoy successor, in 1996. The company's dominance was so total, however, that this 32-bit next-generation

1.

History
of Portable
Devices

Nintendo has often eschewed technical capability in favor of unique hardware that facilitates new kinds of play. The DS touchscreen is a first for a dedicated game system, and the dual screens is part homage to the Game & Watch of old, and part experimentation for Nintendo.

machine didn't hit the stores until 2001, when it was released it as the GameBoy Advance.

The GameBoy Advance was at once a revelation and a disappointment. In spite of an incredibly dim screen that didn't really allow the launch titles to impress, the GBA went on to sell millions of units. Conspiracy theorists suggest that it was Nintendo's fear of Sony entering the portable market that forced them to release the GBA, but Sony remained quiet, and for

several years the GBA continued Nintendo's domination of the portable arena.

However, without any reaction from Nintendo, another portable gaming platform has slowly emerged, one that seems totally immune to Nintendo's ability to dominate the mobile landscape: cellular phones.

Starting with Nokia, cellphones started coming out with larger screens and the capability to play

unsophisticated games. Poorer than the first arcade games seen 20 years earlier, games like *Snake* and *PacMan* were seeing vast hours of playtime from their owners. Although many people wouldn't carry a game system around to get their fix, phone owners found a quick game could kill time regardless of its sophistication.

Between Nintendo and cellphone manufacturers you could go virtually anywhere and play your games. It was a glorious time.

1. Sony PSP, plus UMD

1.

These images of a pre-release PSP show Sony's willingness to depart from Nintendo's vision of portable gaming. Black, glossy, widescreen, and decidedly more adult than Nintendo's hardware.

019

The GameBoy

For many people the GameBoy was the must-have gadget of choice before cellphones and iPods took over our lives.

Nintendo's 1989 GameBoy was exactly the right system at the right time. The NES had cemented Nintendo's position as the new king of the video game, with tens of millions of units reaching homes and pleasing players with games like *Super Mario Bros.* and *Zelda*. With over 40 million Game & Watch systems in hands and pockets worldwide, Nintendo's next move was very easy to foresee. A portable console capable of playing NES-quality games would combine the best of both worlds, pleasing players on the go and delivering the quality Nintendo was becoming known for.

The technology of the time didn't allow for a portable system with full-blown NES capabilities. LCD screens were only monochrome, and NES-level processors were power hogs, which would suck batteries dry in minutes. History has shown time and again that Nintendo doesn't need to be cutting edge to be successful, and the GameBoy was no exception. The quality of the games would sell the system regardless of the system's specifications.

When it was launched in North America in 1989 the screen wasn't as good as other LCDs at the time. The GameBoy's screen wasn't even as clear as the Game & Watch systems it replaced. Instead of black and white elements, the GameBoy screen offered four shades of a color that could charitably be described as algae-green. The screen blurred terribly, and the resolution was much lower than even the NES. The GameBoy could display about one-third of the number of pixels the NES could, but that didn't stop people buying the GameBoy by the millions. In the first three years of release the GameBoy sold over 30 million units.

1. GameBoy Pocket

1.

History
of Portable
Devices

The GameBoy was a massive, phenomenal success. Several versions were released, along with more than a dozen color variations. In spite of its substandard, four-shades-of-green screen it racked up many tens of millions of sales.

The GameBoy's success was built largely on the back of a game licensed from an unknown Russian programmer *Tetris*, the game that inspired an entire genre, was a pack-in title, included with every GameBoy sold. The GameBoy might as well have been designed for the game—the blurry LCD screen didn't exhibit this effect with the *Tetris* blocks, which fell in large steps, rather than gliding smoothly. Players were entranced.

The advertising for the GameBoy was very thorough, and more people were likely to have seen a GameBoy and *Tetris* commercial than any other game before it. People who wouldn't normally play games were addicted to *Tetris*. Thousands of mothers were buying the system for themselves, something unheard of in the gaming world, and all because of a game featuring falling bricks. It had no clever movie tie-in, no recognizable mascot or cartoon character, and no realistic graphics digitized from real-life locations.

And then came *Super Mario Land*, an all-new *Super Mario* game released the same year as the GameBoy, more than two years after the last Mario game came out for the NES. This relative Mario drought, near the peak of the series' popularity, set the stage for an unstoppable portable platform. For the first time ever a portable system was available with games people very much wanted to play.

For the first time ever a portable system was outselling home video game consoles. It was a revelation; people were playing GameBoys on the street, and a phenomenal number of children were playing them in schoolyards.

Mass-market portable pixels had finally arrived, and you could have any color you liked—as long as you liked green.

1. *The Legend of Zelda, Link's Awakening* (GameBoy)

1.

The GameBoy

The GameBoy Pocket gave the original, clunky GameBoy some serious longevity. It featured a higher-contrast screen, and offered four shades of gray instead of green. It lasted forever on two AAA batteries, and offered all the functionality of the original.

The Turbo Express

The early 1990s saw another new kid on the block, which used credit-card sized catridges to attract our hard-earned cash.

Released in Japan as the PC Engine GT, this was and remained the king of portables until Sony released the PSP. It featured an active matrix backlit LCD screen, meaning the response time was blazing fast and you could play it in the dark. It had vivid colors and a higher resolution than any other portable of the time. It cost $400 at release, and for this reason alone it could never have succeeded.

The game was based on the popular Japanese PC Engine or the unpopular TurboGrafx-16, depending on which side of the Pacific Ocean you were on. It featured the same three chips that powered the home console, providing real 16-bit graphics on the move. The Hudson Soft chipset was capable of over 500 colors and larger sprites than its competition, and the screen really showed off the system. Unlike the GameGear,

Lynx, and GameBoy Color, the Express had a screen that could be viewed in nearly any lighting and from nearly any angle. Both the GameGear and Lynx suffered greatly from cheap screens that were very hard to view comfortably, but the Express was very easy on the eyes.

The Express had another ace up its sleeve: the same credit-card cartridges that played on its big brother, the TurboGrafx-16, would play in the Express. For the first time a console was made portable and used the same media. Sega tried this trick years later with their Nomad, a portable version of their dominating Genesis console, but Genesis cartridges were massive compared to the Express' HuCards. A half-dozen Nomad games would fill several pockets, but you could fit a half-dozen HuCards in a wallet.

1. *Super Star Soldier*, Hudson, TurboGrafx 16

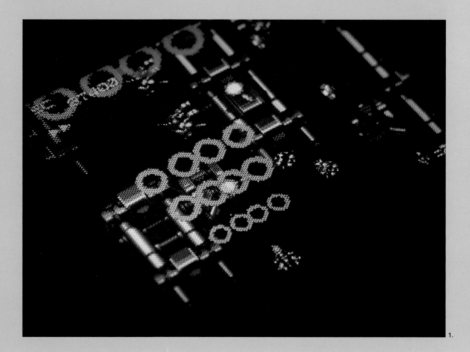

1.

The Turbo Express' screen was second to none at the time. It utilized a battery-sucking active matrix backlit LCD with a higher resolution, faster speed and more colors than any of its competitors.

But there was no hope of success for the Express. It was released too late and it cost far too much for the average gamer. When it hit the shelves it had a price tag almost four times higher than the TurboGrafx console. By 1997 it was being dumped, and you could buy the Express new in bulk for less than $70 each.

1. TurboGrafx 16
2. Turbo Express

1.

2.

The Turbo Express

The Express used the same media as its console brother, the TurboGrafx 16, offering a portable experience that was unmatched. Its price tag, however, exceeded the experience as far as players were concerned. It was a poor seller.

The NeoGeo Pocket

This late 1980s icon was still seeing new content developed some 15 years after its debut.

SNK was no stranger to imitation. Its claim to fame was the NeoGeo, a modular system that allowed arcade operators to swap out one game and replace it with another as easily as changing cartridges in a console. This was a great idea, but in practice it meant that an operator could replace one cheap *Street Fighter* clone with another. The huge majority of SNK's releases were fighting games, and for a long time it seemed they would never be more than an also-ran company with a series of mediocre games. It seemed SNK could not make very good software but could produce great hardware. The NeoGeo system, released in 1989, was still seeing new releases in 2004. No console or platform could claim such a successful time on the market.

SNK was no stranger to the home market either. When the NeoGeo was new, SNK experimented with game rentals in Japan, where it is normally illegal to rent software. While SNK quickly discovered gamers weren't interested in renting the system, it was clear that quite a few people were willing to buy the games—even though single game prices reached more than $300. For some time SNK was very successful, with a string of hit games and a small but fervent group of fans who snapped up every new game. The good times came and went for SNK, and perhaps out of desperation, the company looked for something new and decided to release a portable game system.

The NeoGeo Pocket was an excellent system, both in capability and design. It won awards in Japan for its quality and styling, and SNK pushed the hip new system hard. The advertising campaign featured the slogan "I'm not boy," a sideways shot at Nintendo's GameBoy, which should have been long in the tooth and ready for replacement.

1.

History of portable devices

SNK's NeoGeo Pocket, and later the NGP Color, should have been a success. Its only real competition was the GameBoy, which was bested by the NGP in every respect except one: software. Despite award-winning hardware it never managed to generate must-have software.

1. NeoGeo Pocket Color

2. *Samurai Shodown 2*, NGPC, SNK

1.

2.

The
NeoGeo
Pocket

SNK's reliance on ports of its popular games didn't help the system. SNK was a favorite with the hardcore, not the masses, and while those in the know snapped it up there weren't enough of them. The NeoGeo Pocket continues to enjoy significant collector demand, and few dispute the quality of the hardware.

SNK pillaged its back catalog for releases and for a short time it looked like the company had it made. Everything it had was better than Nintendo's offerings. Games came in plastic cases instead of cardboard boxes and instruction manuals were full color instead of black and white. Several notable third parties created games for the system, including Namco, which released PacMan.

The hardware for the NeoGeo Pocket was solid as well, with a horizontal design much easier for players of all ages to grasp compared to the skinny GameBoy, which cramped older hands. The screen was exceptionally clear and easier to see in lower light and from more angles than the GameBoy. Instead of a flat directional pad, the NeoGeo Pocket had a neat clicky thumbstick that felt great. It ran for

longer than the GameBoy on only two batteries, and had a built-in world clock and horoscope function. Before it was discontinued, SNK released an efficient wireless unit that allowed up to 64 players to play together (although no game supported more than two players). Unfortunately, SNK managed to sink its baby. Almost immediately after launching the black and white NeoGeo Pocket there was talk about

1. *SNK vs Capcom Millenium Fighters,* SNK (NeoGeo Pocket Color)
2. *Metal Slug: 2nd Mission,* SNK (NGPC)

1.

2.

History
of portable
devices

the color successor, due a year later. Few players were willing to pick up the system when another was so close on the horizon. SNK in Japan was notoriously hard to deal with, and no Western game company could even get developer documentation from it. Even Japanese developers stayed away after the initial launch, and soon SNK alone made games for the unit. SNK's American distribution company wanted nothing to do with the unit, and didn't use the fancy plastic cases for their territory or even secure much advertising. Instead, it gave the product half-hearted support, and dumped stock early. At the end, the system was given away as prizes in pachinko parlors in Japan, and Aruze, a gambling machine manufacturer that owned SNK for a short time, released almost ten nearly identical games featuring its latest slot-machine in portable form. Between SNK's inability to keep up the momentum, Nintendo's vice-like dominance, and a lack of great games, the talented little handheld just lost sales. Systems bundled with six games in a bubble-pack for under $50 were available throughout North America only a few years after the system was released.

1.

The Atari Lynx

Atari deserves a place at the high table of computing and portable devices. Its offering was up against some big players in the late 1980s and early 90s.

The Atari Lynx was a console ahead of its time, and though it was the most technologically advanced, it was doomed from the start. A popular game-making company called Epyx designed the Lynx in 1987 but couldn't afford to produce or market it effectively, and sold the design to Atari. It would quickly become apparent, however, that Atari wasn't able to market it effectively either. For one reason or another Atari sat on the completed Lynx design for two years, choosing instead to release it against Nintendo's GameBoy in 1989.

The Lynx was a silicon marvel. Designed by the same team that created the Commodore Amiga computer and later went on to create the 3DO, the Lynx had features no other console did. It had a color screen and was, despite Sega's claims years later, the first color portable game system. It could be played by both right-

and left-handed players—the unit could be inverted, placing the directional pad on the right and the buttons on the left. Up to eight players could connect their Lynx systems and play together, a feat not equaled until SNK's NeoGeo Pocket nearly a decade later. Games came on little wafers called game cards and were much thinner than GameBoy or GameGear cartridges.

Several big-name games were released for the Lynx, including more than a dozen of Atari's biggest arcade hits. Atari's star was fading however, and the company couldn't attract significant developer support. What games Atari could license were typically produced in-house, to varying degrees of success.

1. Atari Lynx with games
2. Atari Lynx

1.

2.

1. *Batman Returns*
   (Atari Lynx)
2. *Double Dragon*
   (Telegames)
3. *Gates of
   Zendocon* (Epyx)
4. *Gauntlet*
   (Atari Lynx)
5. *Klax* (Atari Lynx)

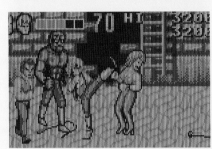

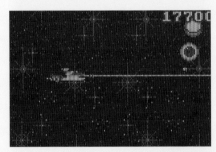

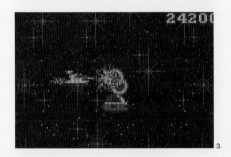

Much of the Lynx library was of better than average quality, and a few, most notably *Chip's Challenge* and *Slime World*, went on to be ported to other systems. Some of the biggest names were instantly recognizable: *Batman Returns*, *Gauntlet*, *Klax*, *Centipede*, and *Double Dragon*.

Despite its superior abilities, the Lynx wasn't able to reach significant sales, and by 1992 most retailers were clearing stock. Support dragged on for another two years, and production of the Lynx finally ceased in 1994.

The Wonder Swan

Later in the 1990s Japanese giant Bandai produced a device that had its own, unique attractions for games designers.

The WonderSwan was a major entry by toy and video game manufacturer Bandai into the handheld game hardware field. The original 16-bit black and white console was released in Japan in 1999 to some acclaim, having been designed by the legendary Gunpei Yokoi, who designed the GameBoy and Virtual Boy consoles. The system was unique in that it could be played horizontally or vertically, and some games made use of this to switch back and forth between various styles of play.

Initial reaction to the console was good, but by now the GameBoy had a color version, and the GameBoy Advance was on the way. So without further delay, Bandai rolled out a color version of the console in 2000, and in the process gained the support of Squaresoft, and a new lease on life, at least in Japan. The color version featured the same

processor but a larger screen and redesigned case. The speed of the virtual RAM was further increased as well, giving the WonderSwan Color the edge until the GBA ramped up. The console never saw the light of day in the USA, but did moderately well in its native country, and in 2001 a final console, the SwanCrystal, was released. The newest system boasted a crisper TFT screen and another case redesign, and was by far the most attractive of the three.

Games for the WonderSwan had an attractively chunky pixel art style and higher color depth than the GBA in the same time period. As it operated on one battery and featured innovative hardware design, the console won multiple awards in Japan during its lifetime.

In a partnership with the small software company Qute, Bandai released the Wonder Witch, an

1. *Puzzle Bobble* (WonderSwan)
2. WonderSwan Color

Capcom (WonderSwan)

amateur development software kit that also included a WonderSwan cartridge on which to play independent games created with the software. Some notable artists, including Kenta Cho and Murasame from ABA Games, contributed games over the years; two titles made it into final production. The last games for the SwanCrystal were released in 2004, and the console has now been laid to rest.

The WonderSwan family of consoles was a niche offering through and through, but through its Wonder Witch service helped to further the development of pixel artists and amateur developers, and as such has left a mark on the gaming world.

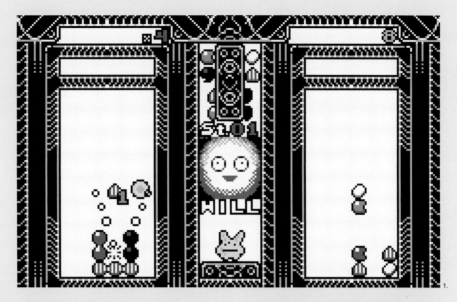

1. *Puyo Puyo*, Compile (WanderSwan)
2. WonderSwan Color
3. *Mr Driller* (WonderSwan)

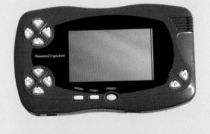

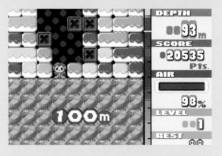

The WonderSwan

Klonoa, Namco (Wonderswan)

The current generation

Just as pixel art and designs influenced by the early years of gaming are becoming massively popular, the technology is attempting to edge out the form in favor of 3D designs.

The current generation of portables is really exploding with fantastic new capabilities. Advances in silicon-chip technology and liquid crystal screens have allowed engineers to pack far more into a far smaller space. You might expect this to be a golden age for pixels, but you'd be wrong. 3D games continue in this new generation to try to edge out pixel games.

These are trying times for pixel artists and game players. While it seems that the pixel has never been more popular, with artists like eBoy and Army of Trolls making magazine covers right and left, the popular gaming attention is apparently devoted to 3D.

Sony's PSP is following the pattern. Sony started with the PlayStation and PlayStation 2, with polygons and realistic-looking games as the focus. Companies like Capcom are remaking their classic titles in shiny new polygons. It would seem that, as far as Sony are concerned, pixels are a thing of the past.

Cellphones too are packing in 3D chips to try and create polygonal worlds and amaze with faster and better graphics. Many Japanese phones are now more 3D-capable than any pre-Playstation console. Japanese gaming giants like Namco and Taito are creating cellphone games that rival the very best console games of just a decade ago. Most phones sold around the world though are still very much 2D, and hundreds of companies are paying pixel artists to create games the old-fashioned way.

1. *Castlevania – Dawn of Sorrow*, Konami (Nintendo DS)

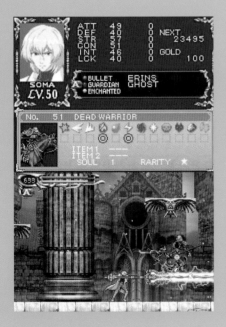

1.

Nintendo's DS, like the GBA before it, is one of the last bastions of pixel gaming excitement. Where mobiles aren't really made for gaming, the DS is, and games like *Castlevania*, *Yoshi's Touch*, and *Go* prove that talented pixel artists still pound out the pixels. Most video-game graphics, be they 2D or 3D, require a dedicated artist and lots of time to create. Pixel art isn't dead, but you need to know where to look.

1. *PacPix*, Namco (Nintendo DS)
1. *Nintendogs* (Nintendo DS)
3. *Zoo Keeper*, Ignition Entertainment (Nintendo DS)

1.

2.

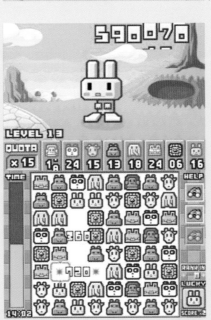

The current generation

Recent history

That brings us up to more recent developments in gaming hardware and software. How have these leaps forward influenced today's character designers and their work?

Nintendo's GameBoy changed everything. It didn't really offer much, especially by today's standards, but it did more than any other portable before it, and was better in many respects than any competitor's portable since. Its drawbacks were obvious: a low-resolution screen that could display only four shades of what might be accurately described as a sickly yellow-green, a slow CPU, and very rudimentary game-specific processing ability. It had stereo sound but was saddled with very poor audio capabilities. GameBoy had Nintendo's trademark directional pad and two buttons, plus Start and Select. It was very similar in function to Nintendo's earth-shattering NES, and indeed made its fame with ports and sequels to many NES games.

It also had *Tetris.*

The GameBoy rocketed to superstardom on the back of Alexi Pajitnov's seminal puzzler, which became the most popular game ever made. The GameBoy was released with Tetris as a pack-in game, and while nearly every computer and console had Tetris, the only way to play it on the move was with a GameBoy. It sold millions, for the GameBoy was ideally suited for exactly this kind of game.

A lot of companies saw Nintendo's GameBoy success and tried to compete, but for one reason or another all fell to Nintendo's underpowered juggernaut.

Atari, still struggling along as a hardware manufacturer at the time, released the world's first portable color system: the Lynx. It featured some very impressive hardware,

1. Nintendo GameBoy Advance

1.

History of portable devices

Nintendo's GameBoy Advance succeeded the original GameBoy and GameBoy Color, and maintained Nintendo's dominance of the portable game space despite sitting on the shelf for six years before release.

a flippable design so left-handed players could use it, and a backlit screen that allowed it to be played in the dark. It didn't have Tetris, or indeed any other recognizable titles, but its bigger flaw was Atari itself, which had neither the money nor savvy to market the Lynx properly. More than 115 games and two different hardware versions were released before the Lynx was put to rest.

The most serious threat to the GameBoy was Sega's GameGear, another color system with more marketing muscle than Atari could employ. When it was released in Japan it immediately developed a reputation as a bit of a junker, and poor build quality plagued it during its entire run. Like the Lynx it would suck six AA batteries dry in less than ten hours. It was unique in its ability to quickly suck dust particles under the screen where they couldn't be removed, and if you pressed too hard on any of the buttons, the LCD

screen would flex and discolor. Although the system was software compatible with Sega's Master System console, it never developed a library of hit games, and eventually Sega gave up on the GameGear.

Before Sega launched its MegaDrive in North America as the Genesis, many in the game and toy industries pegged NEC as the winner of the 16-bit war. NEC's TurboGrafx-16 system was astonishingly popular in Japan where, known as the PC Engine, it had dethroned Nintendo's NES as the system of choice. It didn't win over the Genesis, but it spawned a portable system that was without question the best of its generation. The TurboGrafx used tiny, credit-card-sized cartridges, and when NEC released its color portable, it used the same cards as its full-sized console brother. For the first time you could play your game library at home on the TV or on the move in the TurboExpress. Unfortunately, the high-res color

screen also used up batteries in short order, and this, combined with NEC's disillusionment with video games in general, led to its downfall. The TurboExpress and its exceptional backlit color screen remained the quality king prior to Sony's release of the PSP.

Many others tried to launch successful portable games and failed. Sega even tried again, with a portable version of the Genesis. It had a high-resolution screen and played the entire Genesis library, but it was poorly marketed, Within two years, shows discounted it to one-third of the price. Arcade giant SNK launched its NeoGeo Pocket in Japan, Europe, and North America, but because of a lack of cooperation with developers, its small library never managed to please gamers. Despite winning awards for design it sank only three years after release.

1. Sony PSP

1.

Recent history

The PSP is everything the DS isn't—black, glossy, a fingerprint magnet. Its wealth of features makes it a gadget-fiend's dream machine, though Sony had trouble fitting it all into such a small package.

The Nintendo DS

A recent technology, but a pixel-friendly player that offers designers a massive increase in available color and detail.

Nintendo's DS is the latest in a long line of pixel-friendly systems from Nintendo. It features two screens, one of which is a touch-screen, two processors, eight buttons, and a four-way directional control. When closed, it isn't much larger than the GameBoy Advance, but it's much more powerful. The DS is also the first Nintendo portable system with an illuminated screen from launch, unlike the GameBoy Advance or GameBoy, which had lighted systems released much later.

The DS screens are both larger than the GameBoy Advance screen, allowing for a significant increase in detail, and with a color palette of over 260,000 colors, there are no practical limits to the kind of images artists can create. While the system was initially touted as an equal to the Nintendo64, with all the 3D graphic power that implies, much of the DS library is 2D. It is a machine that is very friendly to pixel artists, offering modern, powerful hardware on a platform that is 2D friendly.

Nintendo's artists have become masters of their pixel craft, creating entire worlds and beloved characters out of the little square dots. On the DS they've had a chance to really shine, with beautiful updates to long-running series, such as *Kirby* and *Famicom Wars* (*Advance Wars*).

Nintendo is not alone with the pixel goodness either. Many companies are releasing games for the DS that feature some of the best pixel art ever made. Years of practice, new techniques, and modern tools make the DS home to some astonishingly cool art, in spite of its diminutive screensize and specs when compared to consoles.

1. Nintendo DS

1.

History of portable devices

The Nintendo DS has also been made available in pink and baby blue to mark the release of Nintendogs and red to celebrate the release of Mario Kart DS.

1. *Nintendogs*
   (Nintendo DS)
2. *Mario Kart DS*
   (Nintendo DS)
3. *Kirby Canvas
   Curse*
   (Nintendo DS)

1.

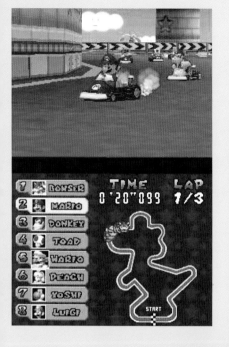

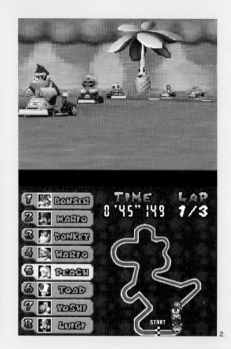

2.

3.

The Sony PSP

As Sony's generation-defining onslaught continues, the portable version of the PlayStation has some unexploited pixel potential for canny designers and artists.

While the PlayStation Portable (PSP) is currently best known as the leader in polygon pushing for handheld game consoles, there is a home for the pixel on this beast as well, if not a populated one. Only a few 2D games make use of the PSP's crisp screen and robust RAM, which both lend themselves quite well to pixel work.

At the time of writing, the grand majority of pixel-oriented games are RPGs, many of them ports of earlier works. *Astonishia Story* is one such port, from Korean developer Sonnori. The game was originally made for the GP32, discussed elsewhere in the book.

Longtime Japanese PC game stalwart Falcom has been one of the few to release original titles for the console, though ports have also been released. One of these new titles, *The Legend of Heroes: A Tear of Vermillion*, has been

brought to the West by Bandai, and is the game we are showcasing here.

The developers took original illustrations and mimicked the style for pixel art busts of the in-game avatars, which appear during dialog sequences and character status screens.

While 2D will likely remain a niche genre on the PSP, simpler graphics have long been an important part of casual games, as well as the more hardcore arcade-style titles. As such, an increasing number of PSP games are in the works that incorporate 2D minigames, in order to capture some of that quirky gameplay. The DS is currently the dominant force in this area.

1. *Ridge Racer* (Sony PSP)
2. Sony PSP

1.

2.

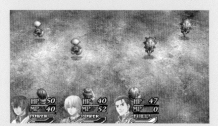

The Sony PSP

Pixel toys

You no longer have to play with pixels solely onscreen. Pixel toys are a tongue-in-cheek reminder of the joys of the humble square when it comes to play.

The pixel has come a long way from its roots in the very first arcade games to consoles and computer games and even to portable games and cellphones. A kind of retro movement has sprung up, making the pixels larger than they need to be to emphasize their "pixelness."

In Japan, where trendy things tend to explode in popularity more than in any other country, pixels have made the shift from the screens and into the hands of the players. Toys and figures from almost every manufacturer have reached the shelves and vending machines across the country.

For example, in Japan you can find tiny dioramas featuring *Super Mario Bros.*, with a bizarre new 3D take on the 2D graphics from the original game. The Mario figure has a magnet inside him, and there is a paired magnet embedded in a coin on the other side of the diorama so that Mario can be moved around on his little stage.

Also to be found are keychains, featuring pixelized creatures from Mario games, captured inside little plastic pouches as if they were fossilized mosquitoes in amber. They come in giant pixel mystery blocks with coin slots on top that can be used as tiny piggy banks.

Similar real-world pixels have hit the mainstream. Pepsi, in a wildly popular promotion, saw a massive jump in sales of their regular and diet Lemon Pepsi when they attached a pixel bottle-cap to every 600 ml beverage.

Pixel construction toys have even been released, with pegboards and square pegs, not unlike the LiteBrite toy of yesteryear, but with a funky modern bent. If Lego was used to create images instead of

1.

objects, the result might have been similar. These popular toys have sold out several times, and feature *Super Mario Bros.*, *PacMan*, *Space Invaders*, *Dig Dug*, and others.

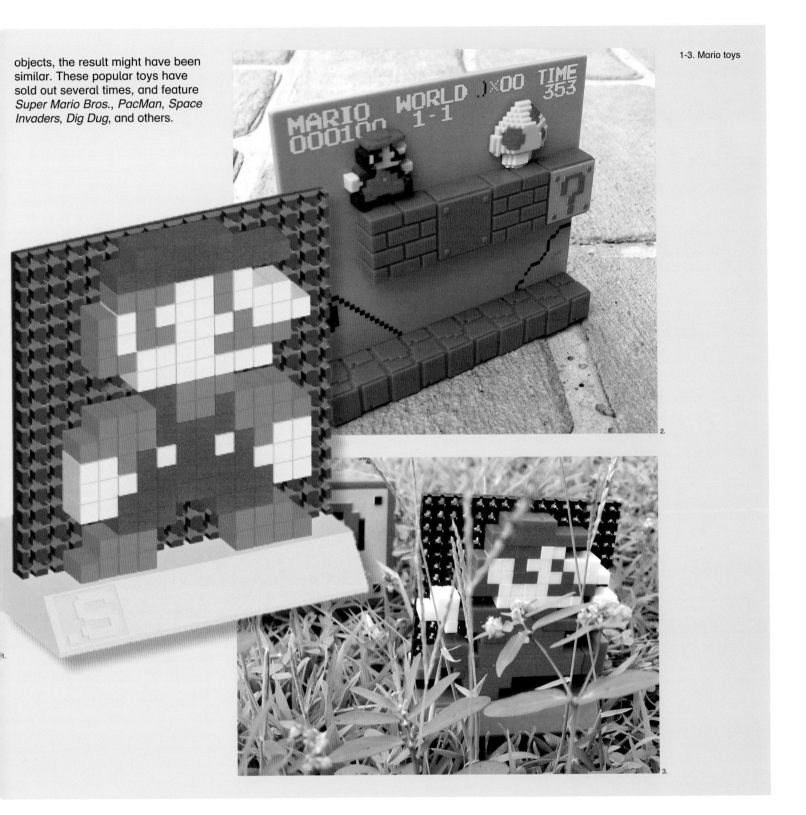

1.

2.

3.

Pixel toys

Theory— changing hardware

The constantly evolving nature of hardware and its capabilities has a massive impact on what designers are able to achieve technically, and artistically.

When video games first came out, they inhabited very primitive hardware. Processors were slow, storage was miniscule, and the development tools didn't exist. Computers were expensive devices, and often the same machine was shared by all the programmers in a company. Graphics usually started as pencil sketches on graph paper and were laboriously transcribed by hand to machine-readable data cards. The hardware they ran on wasn't much more sophisticated. Atari produced an early *Pong* machine utilizing a black and white consumer television in a wooden box, with pieces of colored plastic taped to the screen to highlight scores. Rudimentary they certainly were, and their popularity demanded they progress. One story tells of a bar with a *Space Invaders* machine that had malfunctioned as a result of too many coins filling the inside of the machine. It's no wonder that a color version was soon released.

Color displays preceded decent development tools, and the first colored games were incredibly gaudy, with clashing colors. As few as four colors were displayed at the same time. Component costs were still so high that animated characters were rare, and those that did animate were usually limited to two or three different images. There were blue screens with white dots to outline a race track and cars made of a single color racing around it, but still the quarters flowed.

Things changed quickly during the early Eighties as arcades and the games they offered became an international phenomenon. More companies jumped on the bandwagon, and competition to provide the most amazing thrills was fierce. In 1982, 16 colors on the screen was considered impressive, and artists who had been confined to two or four colors

1. *Tiger Heli*, Taito (NES)
2. Pixel art illustrations of games systems, artist unknown.

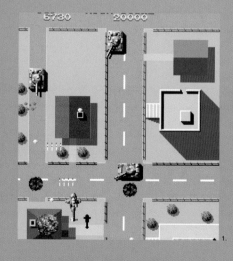

1.

2.

Theory— changing hardware

The platform is always changing, evolving, bettering itself. For every generation of hardware there are only minor differences between competing platforms. More colors, fewer polygons, better sound, less memory. Between generations though the leaps are larger.

started to flex their muscles. Advances in home computers meant that development tools were being created that allowed artists to create images and view them on color displays in minutes, not hours or days.

By 1985 game developers had finally been awarded some incredible tools. Processor prices had fallen so far that many games had more than one. Memory chips used to store graphics were now 2,000 times larger than those used in 1978. The new palette limit for artists was 64 colors, and games were looking better than ever. It didn't stay this way for long.

In 1988 the number of colors onscreen doubled to 128, then 256, and then even more. Storage had more than doubled from three years earlier, and overall resolution was now up to double what it had been in 1980. Using digital photographs, a

technique pioneered as early as 1983 with Midway's *Journey*, became suddenly much more popular with Midway's 1988 release, *Narc*. Using photographs reduced the demand placed on artists, but with the added realism came massive hardware demands. *Narc* was capable of displaying almost 2,000 colors.

By 1996 there were few limits to what an artist could create. Graphics hardware could produce more colors than any artist could conceivably require. The hard part, finally, was dreaming up something worthy of the hardware that two decades of progress had created.

1. *Sengoku Blade*, Psikyo, arcade
2. *Pitfall 2*, Sega, cellphone
3. *Ghouls n' Ghosts,* Capcom, arcade
4. *Last Duel*, Capcom, arcade
5. *Gradius*, Konami, cellphone

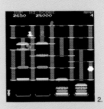

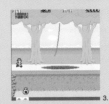

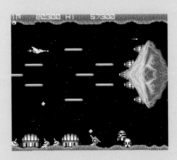

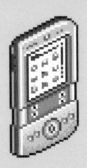

Theory— changing hardware

Compared to consoles the differences between mobiles of a given generation are slight, however they can cause bigger headaches for it. Screens between phones of the same generation can vary wildly in size and resolution; resources and capability are often so different there are more differences than similarities. From a developer perspective keeping up with the spec race is no small challenge.

Cellular devices and games

Many of today's phones are games stations in all but name. How have these developed alongside the great portable gaming devices?

For many of us, portable games have always been around. The primordial ooze from which today's mobile gaming platforms crawled is someone else's history, a story not unlike the tales grandad would spin about "the Great One."

For some of us these early beginnings are our dearest memories. We can remember those early handheld games, those primitive half-mechanical, half-electronic beasts that were always low-tech. We weren't fooled, we knew they were rudimentary, but we also sensed the promise.

Portable gaming is the little brother in the game hardware world. Always a generation behind, always the follower, never the leader, and always wearing big brother's games.

Portable games are rarely as exceptional as their console counterparts. The portable gaming shelves are crammed full of rushed ports, licensed shovelware, and cynical money grabs made to run on hardware that's only noteworthy because it's not attached to a TV.

But still they make these machines, and they make the games, and we buy them and play them by the millions. More and more people are playing games on the go, as games are no longer for the solitary player. They're social. They're short-term. They're convenient.

And every year they get better, faster, more impressive, and these days, finally one might believe that portable games have finally proved their worth.

And one might be right.

1. *Squid* (Animation Test), GameLoft, cellphone

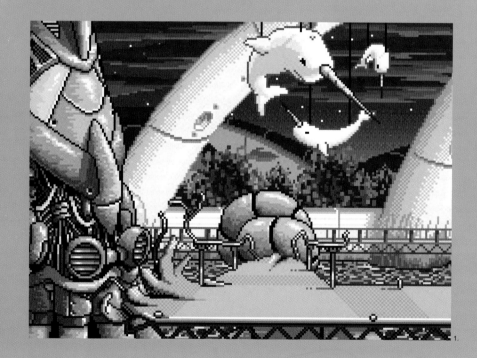

History of portable devices

This is an image test for an unreleased GameLoft cellphone game. Using fewer than a hundred colors, it engages the animation, and conjures up a wealth of backstory. What's a giant robot squid doing here? Are those narwhals? Why are they hanging from strings?

Mobile games are a burgeoning industry at the moment, though some suspect that the proverbial bubble may burst. Because the hardware platforms, that is to say the phones themselves, are still in their formative years technology-wise, and also because of the nature of the tiny devices, the pixel has found a new home here. Though the gameplay experience on cellphones still often leaves much to be desired, the art can be incredibly charming.

Many liken the current mobile game environment to that of the classic console days, with smaller teams, smaller budgets, shorter development cycles, and bright-eyed ideas about potential for the future. The big difference though, is the huge number of devices these games must be ported to, the data constraints with which artists must work, and the types of games that actually sell.

Some lament that even in the mobile world, where 2D is currently king, the polygon will push the pixel out of the limelight soon enough. But the appeal of retro gaming is strong here, and you can't do retro without the pixel! And while the games may not fit all tastes at the moment, with companies such as Square Enix, Namco, Capcom, and Hudson

making significant moves into the mobile arena, it's only a matter of time before a compelling "killer app" shows up on the radar.

1. *Lord Monarch*, Falcom, cellphone

2. *Prince of Persia*, GameLoft, cellphone

1.

2.

History of mobile devices

These images illustrate the diversity of styles and visual richness available in modern cellular games. Mobile games now often run on more capable hardware, with color and storage designers only dreamed about in the 1990s.

Cellular gaming

Cellphones have gone from being luxury items to must-haves for millions of people. But what are the issues when it comes to designing for tiny screens and diposable hardware?

Cellular phones have created a whole new way to play games. No one could have predicted how little time it would take to move from the first shoebox-sized cellphones to the slimline phones of today. The modern cellphone has placed an astonishing amount of power in the palm of anyone's hand, more power than entire computers less than 15 years ago. Cellphones offer better-quality screens capable of more colors than any portable game system excluding the very newest, and they are very cheap.

It's not surprising that cellphone users want to do more with their phones. Now they can use the built-in camera for capturing images, and Bluetooth and other wireless technologies for sharing them with other cellphone users. Fast 3D chips, TV and stereo sound output, flash card slots and bar code scanners, and all kinds of fun toys are part of the package.

Like any other maturing technology the attention turns from function to entertainment. Many people are playing games on their phone more often than making calls. Game developers are taking advantage of this new platform, creating vast quantities of new games. Pixel artists are once again in demand as graphics and games are crammed into tiny screens.

There are hundreds and hundreds of cellular phone games available to many phone users, but despite voracious demand and developer willingness the cellphone is not a very good games platform.

The control pad on a cellphone is terrible. It's unresponsive, stiff, and almost totally useless for action games. It's nigh impossible to move the average thumb across a flat disk accurately at speed, and as every player knows, if you can't control your game, you die.

1. Toshiba J08-T, J-Phone / Vodafone

1. *Die Hard*, Mobile Scope (C64, NES, TG16)

1.

2.

Theory—changing hardware

Phones are always changing, and especially in Asia the manufacturers are always trying to one-up each other. The Toshiba shown above had a higher resolution screen than the competition, but it couldn't run high-res games. The strain of enlarging low-res games crippled the processor, rendering most games unbearably slow.

The action button, operated with a second hand in almost every console game controller ever made, is now in the center of the directional pad on a cellphone. It's incredibly difficult to push the button at the same time a direction is being pressed, so the number keys must be used instead. Many games, in fact, offer the number pad as a replacement controller, a control method that was common in computers more than a decade ago. It's not an ideal control scheme, but it works well enough that phone users download hundreds of thousands of games every year.

1. *Die Hard*, Mobile Scope (C64, NES, TG16)
2. Sharp V601SH, Vodafone
3. *Die Hard*, Mobile Scope (C64, NES, TG16)

Cellular gaming

The Sharp phone in this image was the last of the 2D phones. It could run every game that came before it, but was king of the hill for a mere months before 3D models were passing it by.

How cellphone differs from console

What are the challenges of designing sprites to play on cellphones and similar devices?

Creating graphics for a portable platform, particularly cellphones, involves a very different set of criteria than creating for game consoles. Until recently, the capabilities of portable systems lagged far behind those of consoles: fewer colors, fewer sprites, less hardware-level support, and much less storage. For a very long time, this meant that portable graphics were generations behind, and pixel artists creating images for these new platforms had to utilize old-fashioned skills.

It is decades since consoles and computers struggled to overcome four-color palettes: CGA graphics for IBM PCs came out in 1981. The GameBoy also had four shades in 1989, though they were all green. Almost 20 years from the CGA days, and eight years after the GameBoy, Nokia released the two-color 6100 series cellphone, which came with the game Snake. Taneli Armanto, who created the game for Nokia, could not have known how much of a revelation it would be. It was one of the first phones to come with features not directly related to the business of phoning people.

In 2000 the majority of phones for sale had color screens, at least in Japan. By 2001 black and white phones were rare, and by 2002 even the prepaid semi-disposable phones all had color screens. At first they didn't have many colors— 256 colors was the best they could manage in 2000—and the saturation and contrast were both poor. The resolution was low as well; early color phones were considered advanced with a mere 120 x 140 pixels, not even on par with the GameBoy, which was more than a decade older.

1. *Snake*, Taneli Armanto (Nokia)

1.

When cellular phones had capabilities similar to decades-old consoles, the techniques used for making graphics on them also tended to be decades old. Every time there was a new generation of phone, which happened with startling speed, there were massive increases in capability. These increases allowed artists to utilize more and newer techniques and tools to create graphics for them.

Nevertheless, the same concerns are present. The storage required for images is the single biggest factor in the size of cellphone downloads. Where the average GameBoy Advance game is now 4MB, the largest cellphone game tops out at 256KB—one-sixteenth the size. It is no easy task utilizing hundreds of thousands of colors, creating fluid animation and a variety of enemies, and cramming them all into a tiny

space that console developers would have mocked a whole decade earlier.

1. *Moorhen Seasons*, Mobile Scope

How cellphone differs from console

The vertical orientation of the cellphone screen is the biggest challenge for many developers. Sideways-scrolling games cannot offer the larger, panoramic view of a horizontal screen, reducing visibility and increasing difficulty.

Physical limitations of LCD

What are the technical plusses and minuses of having LCDs as your main display medium for your character designs?

Excepting a few pre-GameBoy systems, all portable games use liquid crystal displays (LCDs), as they have done for a very long time. LCDs are cheap, capable, and reliable, but they're not perfect. They are not as adaptable as cathode ray tubes (CRTs), which have been used in computer monitors and TVs for decades. LCDs blur, they break, and they have bad pixels that refuse to turn on or off as required. As they get larger, they become more expensive, not just because of the size of the screen but because of the components that drive it.

Off-axis viewing is poor too. If the viewer is not directly in front of the screen the image will change, becoming darker or changing color. In extreme instances, moving too far away from center will invert the colors entirely, resulting in a negative-color image. Saturation—the ability to create vivid color—is

also poor in LCDs compared to CRTs. Before the Sony PSP, no portable screen could display colors very well at all.

For all of these flaws, LCD screens are not likely to go away any time soon, as they have no real competition. The readiest alternative is organic light emitting diode (OLED) technology, currently in use in some portable music players, cellular phones and other niche applications. No other display technology has yet reached the market even approaching the same performance and cost as LCDs.

LCDs do not create light but rather block it, so every LCD screen requires a light to function. The first black and white LCDs used environmental light, such as the sun or a lamp, to illuminate the display. This worked well for a time, but color LCDs require much more light and can't rely on

1. WonderSwan

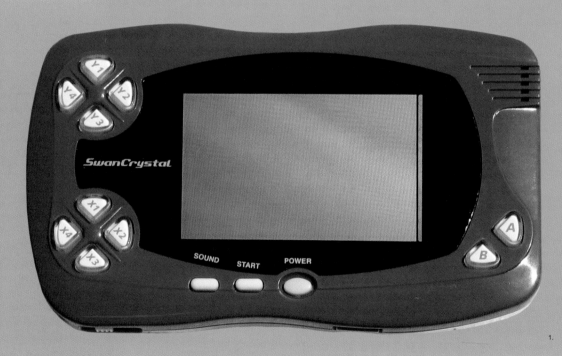

1.

Theory—changing hardware

The first GameBoy had very poor contrast, and only offered four shades of green to the player. More modern consoles, like the WonderSwan, offered more of everything, but even the best LCD is still no match for a good TV. Of course, there's the problem of squeezing a TV into the average pocket...

external sources. When the Lynx, GameGear, and TurboExpress were released, they used backlights—expensive and power-hungry fluorescent tubes tucked behind the screen. These provided ample light but drained six AA batteries in four to six hours.

Modern portables such as the GameBoy Advance SP and DS use a different kind of lighting: frontlights. A special plastic film is placed on the LCD screen between the screen and the player. Lights at the edges of the screen direct light against the edge of this film, and this light is reflected downward, into the LCD. This redirected light passes through the color LCD and reflects off a silver layer at the bottom of the screen and into the player's eyes.

This is almost as ideal a method as LCDs can achieve, though there are two major disadvantages using frontlights. The light is slightly brighter at the edges than the center,

and the entire LCD must be illuminated, including the parts that are black. This is obviously wasteful, but there is no way to target the light only at the pixels that are not black. It is for this reason that both front- and backlit LCDs cannot display a true black. LCDs are not able to block 100 percent of the light passing through them, so in a dark room an LCD screen displaying solid black will still emit a significant glow.

One of the great advantages of CRTs is their ability to dynamically stretch an image on the screen without using expensive CPUs for the job. By expanding the vertical or horizontal size of the image, a CRT creates a bigger but no less accurate image. Many older games took advantage of this effect, easing the CPU load by creating smaller images and letting the TV stretch the image to fill the screen. LCDs cannot do this. Any image stretching must be done by the CPU, and since it's an intensive process, the alternatives

are to create a smaller image—not acceptable on an already tiny LCD screen—or slow the game down to allow the CPU to keep pace.

LCDs are slow to react, so moving objects blur, often quite badly. Recently developed techniques help minimize the effect, but no LCD is quite as good as even an average CRT. Contrast isn't very great, and saturation is lower than most CRTs. But there are no alternatives to LCDs yet, and it appears that the drawbacks of LCD screens don't really impact on players very much. Recent portable systems have made great leaps in CPU power and battery life, and the screens are clearer and more capable than ever before.

The little LCD is tried, true, and able to keep up, no doubt for as long as we need it.

1. *Rally-X, Ridge Racers* loadscreen, Namco (Sony PSP)

2. *Super Star Soldier*, Hudson (TurboExpress)

3. *Slime World*, Atari, (Atari Lynx)

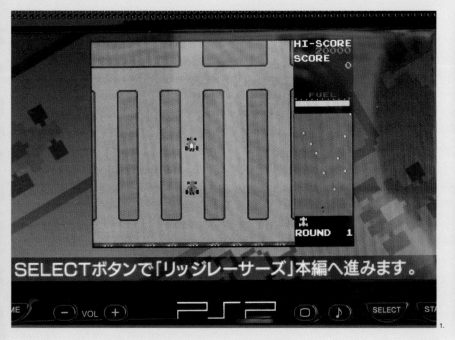

Physical limitations of LCD

These three screens cover a significant range of LCD development. The PSP offers amazing color, and a widescreen format, but isn't very battery-friendly. Compared to it, the TurboGrafx Express had a grainy, blurry screen, but even this screen was worlds better than the Lynx, whose ultra-low-res screen was difficult to see under any conditions. Both of these older platforms would suck six AA batteries dry in under six hours.

Pixel blurring

Designing for any mobile devices means a tradeoff between quality and accessibility and playability when it comes to making your characters move.

Blurring has long been the bane of LCD screens. Moving images on an LCD screen tend to do very odd things, with both the leading and trailing edges blending into, or often "eating away" the background. This drastically restricts the speed and size of moving objects on an LCD screen. In older systems it made the graphic quality of many genres essentially irrelevant as, once in motion, objects were completely unrecognizable.

The original GameBoy's screen rendered any moving object as a shapeless blob if it was small enough and moving fast enough. *Super Mario Land*, one of the flagship GameBoy launch games, was virtually a game of dark jumping blobs. The game could still be played, but the vibrant details of Mario on the television screen were completely absent. Some GameGear games were so

bad that players would die without seeing what killed them, and sometimes without even realizing there was anything on the screen that could threaten them.

Modern screens still have to make tradeoffs in quality, capability, and price. Sony's PlayStation Portable has a screen with the most brilliant saturated color of any portable game system, but it suffers from

1. *Tetris DX*, Nintendo (GameBoy Color)
2. *Tetris DX*, Nintendo (GameBoy Color)
3. *Rally X*, *Ridge Racers* loadscreen, Namco (Sony PSP)

1.

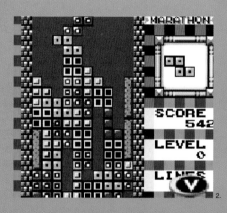

2.

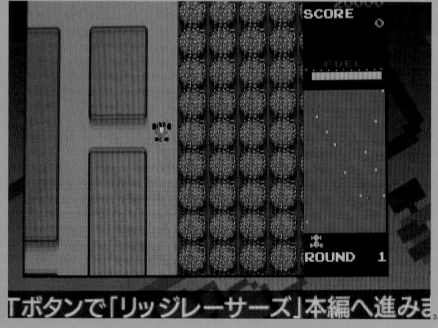

3.

Theory— changing hardware

Pixel blurring is one of the most significant problems with modern LCD screens. Great advancements have been made in reducing the effect, but it's still a problem. The Sony PSP suffers more than most, an intentional trade-off made to increase the glorious color capabilities and reduce costs.

blurring. Moving objects dim and develop large, discolored borders as they shift across the screen.

Motion blurring isn't limited to LCD screens. Televisions and monitors exhibit the same effect but in a different way and to a greatly reduced degree. It's most noticeable in a darkened room, with high contrast objects. Stars in the blackness of space will blur noticeably, but unlike on an LCD display, this blurring doesn't make the moving image significantly more difficult to watch. Whereas a television monitor pixel can activate instantly and fade slightly less instantly, an LCD pixel needs more time to become both lighter or darker.

Liquid crystal is slow to change color. Consider a white square moving across a black screen—on a monitor, this square is illuminated more or less instantly. The pixels behind the square take a short time to return to black, but black pixels will fully illuminate immediately. On an LCD, any changing pixel takes time to change. Black pixels in front of the square take time to alter to white, and pixels behind the square take time to fade. The result is that the leading dims, the black

background seems to resist the movement of the square, and the square becomes a rectangle as both the leading and trailing sides are not fully illuminated.

Clever texture selections can minimize the effect—if the difference in color between two moving objects isn't great there will be minimal blurring. Many 3D PSP games take advantage of this effect. Consider *Ridge Racers*—most asphalt-colored pixels are adjacent to other asphalt pixels, most sky pixels next to other sky pixels. These pixels will change very little from one frame to the next, so there isn't as much blur as might be expected.

Pixel blur can be almost completely eliminated in some games. Card games, RPGs, puzzlers, and other

slow games can mask the effect simply by not having pixels change very often. Even *Tetris*-like games have pieces that don't glide across the screen so much as jump, so the leading-edge erosion is completely negated.

Newer screens blur far less than older ones, and even the worst modern LCD is much better than the earliest ones. LCD computer monitors have pixel-change response times that rival, and sometimes better, the pixel fade times of a CRT monitor. It won't be long until further advances in technology allow portable games to offer screens that surpass CRT monitors.

1. Pixel Blur
2. *SNK v Capcom Card Fighters Clash*, SNK (NeoGeo Pocket Color)

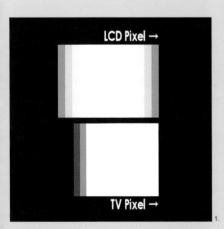

1.

2.

Pixel blurring

A CRT will brighten a pixel to full strength in a single pass, one 60th of a second, but LCDs are as slow to brighten as they are to darken. High contrast moving objects on a TV may leave a trail, but the leading edge of the same object on an LCD will also be eroded as the screen struggles to keep up.

Resizing pixel art

Resizing character sprites is no easy task, as it involves adding pixels or removing them from your already minimal design.

One of the most time-consuming parts of the conversion process is image resizing. Sprites don't change size gracefully—the size of a screen pixel cannot be persuaded to change, so new pixels must be added to or removed from the source image. This is laborious work. If each pixel of a sprite cannot be enlarged by an integer, then it will become uneven as some lines or rows double in height and some don't.

Consider the character Megaman. In the first image below, the original Megaman sprite is on the left. If this sprite is to be used on a screen that's 25 percent larger, the sprite must be enlarged. In the next image Megaman has been "hard scaled," doubling every fourth pixel to achieve 125 percent his original size. This has resulted in some odd effects—his mouth has doubled in vertical size and now appears open, and the black border has new, unevenly distributed black blobs.

Even at 150 percent the effects are not attractive. It's not until Megaman has exactly doubled in size, so that each pixel is twice as high and wide, that the sprite's normal proportions are regained. Any sprite that is scaled up in any uneven increment will need extensive reworking, to the point where in many cases it's quicker to redraw than fix.

Clearly this approach doesn't work well, so a "soft scaling" approach is attempted. Instead of simply increasing the size of certain pixels to achieve the desired result, the image is stretched, and whenever a pixel covers more than one new pixel, it is averaged with its neighbor. In image 2, the pair of stripes on the left has been enlarged horizontally by 50 percent, creating a new stripe that averages its neighbors. When this image is further increased by one-third, the center stripe is replaced by two averaged stripes, and the original image is lost. As you can see, this soft-scaling approach, which

1. *MegaMan*, Capcom, hard scaling 100–200%
2. Pixel resizing
3. *MegaMan*, Capcom, soft scaling 100–200%

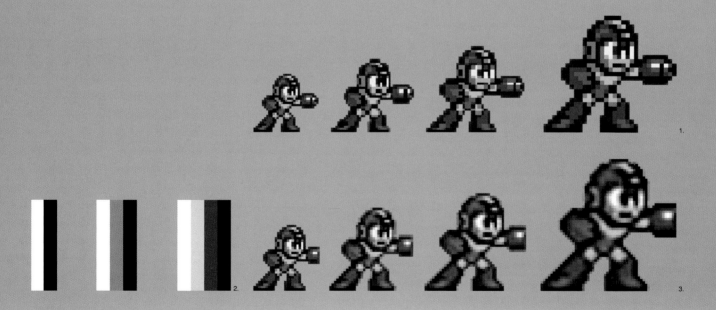

Theory— changing hardware

In the image above, a pair of bars (one black, one white) is enlarged. At a 50 percent increase in size the area between them becomes blurred, a gray line approximates the two colors. Further enlarging this image results in four separate shades of gray—the two bars in the center are trying to blend the white and black bars with the new, gray, center bar.

resamples and averages the pixels, does a much better job. It is most effective when enlarging photographs, which can be enlarged significantly without distorting the image since the real world isn't as visually exacting as sprites. For sprites it just doesn't work well—every pixel was carefully placed for maximum effect, and soft-scaling completely undoes the artist's work. In addition, it requires many more colors to create the new, averaged pixels. To produce the largest Megaman image from the smaller one in image 3, the colors used increased from 13 to 1,407—almost every single pixel is a unique color. This is the underlying hazard when letting the computer handle pixel work: control is lost, and the image is no longer able to conform to the requirements of the target platform. As part of the soft-scaling process, the outer pixels are averaged with

the background. This only works when the background is available, which is impossible if the sprite is being created before the game is played. Blending the resized sprite with the actual game background as it's being played creates an enormous amount of work for the CPU. An alternative is to scale the sprite against a background that is

similar, on average, to the background, but this gives it a subtle aura, or border. If the sprite is taken out of this background, for example out of a green and onto an orange background, the aura can be severe, distracting, and ultimately ugly. Several techniques for overcoming this will be discussed throughout this book.

1. *MegaMan*, Capcom

1.

Resizing pixel art

This image of MegaMan was created on a green background, then enlarged. In addition to inheriting many new colors, each approximating its neighbors, MegaMan grew a green aura. As long as he stays on the original background this aura is beneficial, blending him into the background, but when he's moved onto a different color the new aura is a limitation.

Creating pixel art

How do pixel artists and designers approach their work when the tools at their disposal are constantly evolving?

Pixel art is a modern digital handcraft. It is tedious work, and there's very little that can be done to speed up the process. Most efforts aimed at increasing the speed of pixel art creation tend to introduce effects that are, at best, less attractive. Creating 2D art by pre-rendering 3D polygons (*Donkey Kong Country*) or digitizing real-world designs (*Mortal Kombat*) are both techniques that speed up art creation at the expense of esthetics. Computers can't be trusted to put the right pixel where it'll look the best, and touching up a series of digitized or pre-rendered art is as time consuming as drawing it from scratch.

Earlier artists had more excuses for poor-quality art than modern artists. Primitive tools made their job very difficult, there were no personal computers or graphic design software packages, and often graphics were created on paper and transferred to the hardware using the keyboard to type the value of each pixel. It was often difficult to determine in advance if graphics would look good or animate well before completing the creation and programming phases of design.

After more than 30 years of experience in producing games, pixel artists are better able to wield advanced skills and tools than their predecessors could have imagined.

Modern tools eliminate much of the guesswork and tedium, but not the need for artistic skill. The capability of the hardware certainly goes a long way toward allowing more colorful and detailed images, but the skill of the artist remains paramount. Given a canvas, a brush, and some paint, an artist with practice and education will probably produce better results than someone who has never painted before.

From the very earliest arcade games, it was obvious that some companies had artists capable of working within the medium and some did not. The urge to make a quick buck overwhelmed the artistic sense of many game companies, a practice that continues to this day. Games that don't measure up to their peers are commonplace, but so are games that don't measure up to any artistic standard. Some of the worst offenders are clones—games that are copies of another game, sometimes but not always tweaked just enough to keep the copyright lawyers at bay.

The screenshots here show four very similar games with wildly different graphics. Even though they are all based on the same *PacMan* hardware, *PacMan* exhibits a better design than the other three. Caterpillar is perhaps the best of the rest, with recognizable (though unattractive)

characters. *Joyman* has a maze so poorly constructed that the main sprite, a pulsating abstract thing, passes through some of the walls. *Piranha* is plain ugly, with strange, unrecognizable backgrounds and very poor characters.

As with any medium, practice makes perfect. Pixel art has some inviolable rules. First is the granularity of pixels; details cannot be squeezed between pixels later if they're forgotten early on. Modern computers don't have color or palette limits, but cellphones, portable consoles, and older hardware do. Choosing the proper tool is critical, as many paint programs do not allow the kind of control pixel art often demands.

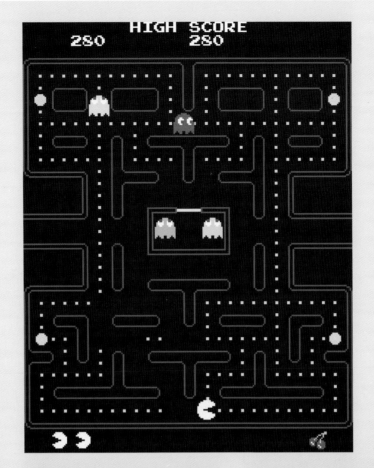

1. *PacMan*, Namco
2. *Joyman*
3. *Piranha*, GL
4. *Caterpillar*, hacked by 'PHI'

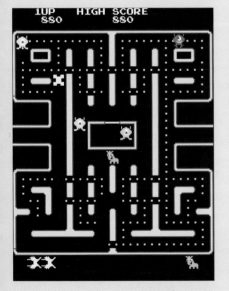

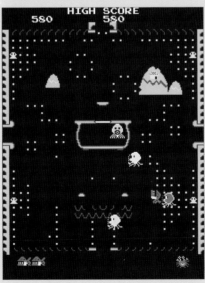

1.

Creating pixel art

All four of these games were created on identical hardware, but only one of them has a really coherent style. The other three are ugly, and appear unfinished or even broken.

Choosing the right tool

So what are the tools of the trade for the pixel artist, the character designer—and the aspiring pro?

Pixel art, like any kind of art, requires the right tools. Not all art programs have equivalent abilities; modern software especially tends to eschew strong single-pixel precision in favor of mass-pixel editing. Adobe's Photoshop, for example, works wonders on photographs but is inappropriate for pixel art, and in fact doesn't work as well for the task as some pixel-focused programs. Despite this different focus and a steep price tag, Photoshop is nearly ubiquitous and very popular with pixelers.

When selecting a program in which to draw pixel art, it's important to keep in mind what your needs are. For many artists, it can be worth the time and effort to seek out older software, that specializes in pixel art. Many modern pixel artists maintain old hardware with the express purpose of running old graphics software on it. The Amiga is an all-time favorite with pixel artists, though some maintain even older hardware, such as the Atari ST or Commodore 64, to achieve certain styles. Many pixel artists enjoy the challenge of creating art for older platforms: it is a very specific challenge to create meaningful, recognizable images with only two colors for every 8 x 8 pixels, or restricting an entire image to 16 colors.

Even without going through the effort of dusting off old computers and trying to integrate them into a modern networked environment, there are plenty of options for pixel artists. Hobby programmers making games for older hardware are always having to create programs to make the artist's job easier. While working on the TurboGrafx-16 game PC Gunjin, artist Ian McPherson had to create whole background images, then laboriously cut them up into 8 x 8 pixel blocks, reduce the colors of

1. Autodesk Animator

INI   FLIC   PIC   CEL   TRACE   SWAP   POCO   EXTRA

Fill

HOME	UNDO	REDO	↑ ←	1	→)) ↓	T F K	FILES...	·	
DRAW	BOX	ZOOM	C	PAN		OPAQUE		V GRAD
POLY	TEXT					GLASS		SCRAPE
SPRAY	FILL	MASK		GRID		SOFTEN		TILE
LINE	MOVE	A				UNZAG		JUMBLE

1.

Theory—changing hardware

Autodesk Animator used to be the most advanced animation tool available. As the latest in a long line of graphics software, its combination of tools and abilities was unparalleled. It still does many things better than modern tools, despite its now-obvious limitations.

each block, and save them as individual pieces. To make the job easier, a program was written that could do the same task automatically.

Custom programs don't always need to be written. There are many pixel editing applications that offer features tailored to the pixel artist. One such program comes from a German group of demo coders. When no other paint program would meet their needs, they wrote their own. Called Grafx2, it's an excellent program that offers many features specific to pixel art. It accurately captures the feel of old-time programs for the Amiga and Atari ST, including the maximum resolution of 640 x 480. Unfortunately, it doesn't offer many modern features and doesn't utilize the normal Windows UI.

Another excellent application comes from Japanese developer Human Balance. Their graphics app, called Graphics Gale, has proven to be so popular it was made available in both Japanese and English. It offers a plethora of graphic tools normally found in high-end graphics applications, such as transparent layers and animation. It fully supports most modern graphic formats, is completely customizable, and surprisingly easy to learn.

For the more adventurous, an excellent MS-DOS application called Autodesk Animator is a fantastic way to generate retro pixel art. It offers full GIF support, with 256 colors and an incredible set of animation tools. It is the final evolution of a set of programs that began on the Atari ST as Aegis Animator and CyberPaint, each new release bringing new and more powerful features to the user.

Unfortunately, it doesn't run under Windows, and requires a DOS emulator such as DOSBox to function.

1. Adobe Photoshop
2. Corel PaintShop Pro
3. Grafx2
4. Graphics Gale

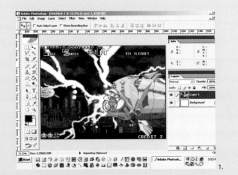
1.

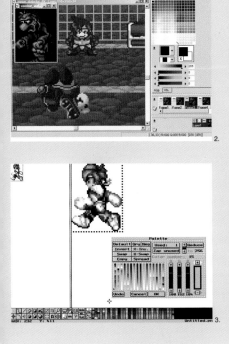
2.

3.

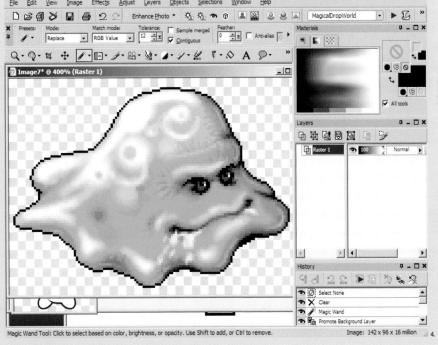

Magic Wand Tool: Click to select based on color, brightness, or opacity. Use Shift to add, or Ctrl to remove.	Image: 142 x 96 x 16 million	4.

Choosing the right tool

For pure pixel art Graphics Gale is the winner, but no program does it all. Most artists maintain a stable of graphics tools, each specializing in one way or another.

Artist focus: Henk Nieborg

Principle artist Henk Nieborg has made a name for himself by creating a signature style that stands out in a world of copyists and clones.

Henk Nieborg has been pushing pixels for a very long time.

Back in 1985, as Henk says, he could be found pixeling with his joystick on a black and white TV. Since he couldn't afford his own color TV, Henk would memorize the Commodore 64-color palette, draw the images in black and white, then check them on the color TV in the living room. He used to make pixel art in the early days with Koala Paint and a Suzo joystick.

Henk has been the principle artist on many games, all featuring a distinct style, with incredibly lush backgrounds and very detailed sprites. From the very first efforts with the Commodore 64 to his first commercial relase on the Commodore Amiga, Henk constantly worked to improve his skill and the results are unparalleled.

Ghost Battle was the first published game for Henk, and it garnered considerable praise. It also landed him a full-time job with Thalion, an intensely popular but short-lived German publisher of computer games. His next game was Lionheart, which sold well considering its platform, but not well enough to keep Henk employed with Thalion which, like most publishers of the day, peaked early and soon folded.

The Misadventures of Flink was next, released for the MegaDrive and MegaCD. It featured some of the most impressive graphics ever released for the MegaDrive, with backgrounds and sprites that did astonishing things with the relatively muted palette offered by the hardware. Flink looked great but was released too late in the MegaDrive's life. It wasn't supported well by its publisher and it achieved unremarkable sales.

1. Art from an unreleased mobile game

1.

This image from an unreleased mobile game shows a facet of Nieborg's artistry not often seen. There are elements that closely resemble several other Nieborg games, but the final effect is almost reminiscent of *Metal Slug*.

The game did, however, land Henk another deal: *Adventures of Lomax*. Released by Psygnosis for the Sony PlayStation, Lomax continued a trend for Henk—games released on platforms that were no longer appropriate targets. Sony had convinced the world that polygons were the future and that sprites were old-fashioned, so Lomax didn't sell well. The press ignored this pixel relic released in a bold new polygon world.

Henk shifted his focus to the portables, which continue to be the last place to find significant amounts of pixel art. First with Disney's *Atlantis* for the GameBoy Color, and later with *Harry Potter* for the GameBoy Advance, Henk's pixel prowess finally found an appreciative audience. And from there? When he's not pixeling, Henk can be found working on textures for 3D games such as *Harry Potter* and *Batman*

Begins. 2006 found him working on a new game, a big release for the PlayStation 3 and PSP that was, he said, a closely guarded secret.

1. *Ghostbattle*
2. *Ghostbattle*
3. *Misadventures of Flink* sprites
4. *Misadventures of Flink* sprites

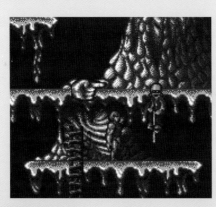

1.

2.

3.

4.

Artist focus:
Henk Nieborg

The top screenshots are from one of Henk's first works, while the sprites beneath them are from the *Misadventures of Flink*, where his style and abilities had become much more mature, elevating him above many of his peers.

Asked about the popularity and future of pixels, Henk remains philosphical about their relative unpopularity. "Pixel games on the big screen could work both ways. More exposure means a bigger market, which in my opinion kills a lot of the creativity. Just look around at what's happening now with the big game industry. Keeping it small and special means people will appreciate it even more."

Regarding pixel art's resurgence in portable games and even outside of gaming, Henk thinks, "Pixeling is almost regarded as a real art and I think it really deserves it. There will always be pixels one way or another, profitable or not. You still need pixel skills for cellphone, GBA, and Nintendo DS games. Even on next-gen platforms, pixel skills are suitable for HUD and icon graphics."

On portables
"Most portable gaming is nothing more than a portable version of a console that was already very successful a couple of years previously in another form. For instance, the GBA had a lot in common with the SNES, and the PSP is actually a heavily modified PS1. They're all great platforms with huge potential, but no one really took the opportunity to make something really nice on them."

On cellphones
"I'm really happy with cellular phone game development because it opens up a lot of opportunities again for pixel artists like me. The quality of games is getting better and better but there's still a lot of mediocre stuff around. This is also a good development because there might be new talent out there struggling to get into the gaming scene."

1. *Harry Potter*, Electronic Arts (GBA)

2.

Theory—
changing
hardware

This background, from Electronic Arts' GameBoy Advance version of *Harry Potter*, evokes a very real sense of place. With very few colors and a limited resolution, the image looks as if it were a representation of a real greenhouse somewhere.

Henk describes the process of developing Flink: "I think there were a few reasons I started drawing everything on a black background, not just on the MegaDrive. First, back in those days I drew most of my graphics on a black background—I just liked that. You could get away with a lot of stuff by fading it into darkness. It's also quite handy when you're dealing with systems like the MegaDrive that didn't give you much memory to play with. I also prefer to draw to blackness because the contrast on the MegaDrive system was insane. If I'd have anti-aliased everything to white I would probably have gone blind."

1. *Harry Potter*, Electronic Arts (GBA)
2. Diagon Alley, from *Harry Potter*, Electronic Arts (GBA)
3. *Misadventures of Flink* sprites

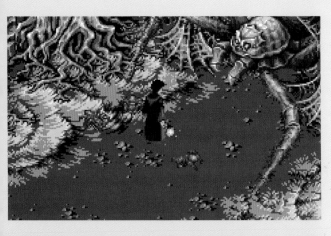

1.

2.

3.

Artist focus: Army of Trolls

For "trolls" read sprites, and for "army" read one man with a mission to populate the world with pixels.

Army of Trolls is Gary Lucken's one-man army. The Trolls themselves are Gary's little sprite/icon characters, which he creates in a somewhat random fashion from doodles on mounds of paper throughout his house. Sometimes the trolls are put up for adoption, and they find loving homes on other sites. As Army of Trolls, Gary has also done some very large and popular works for magazines such as *Edge*, *MacFormat*, and *Webdesigner*. Gary covers a lot of ground, working on everything from massive isometric cityscapes to game sprites and tiny icons.

All of the Troll Army's pixel art is done at 72 dpi. On occasion, he'll double up the pixels to 200 percent: "I quite like the way this looks sometimes, but the majority of the time I'll do the image to 100 percent the desired size at 72 dpi." The Army's weapon of choice is Adobe's ImageReady, which "has all the elements of Photoshop but is restricted to low res, which is perfect for pixels, plus the animation tool is easy to use for testing game animations."

Gary says, "There is definitely a pixel renaissance happening, which is probably down to a few things. A lot of people who grew up with video games are now in their twenties and thirties, so there's a large consumer base for this kind of art. It has a real nostalgic feel to it, which is very unusual for a fairly recent art style. Combined with the growing market for cellphone games, pixel art is a viable career move again for budding artists."

Theory—changing hardware

Army of Trolls produces the extremes of pixel art, from massive spreads to tiny pixel icons and characters. The black cat above is often hidden or tucked away on larger works, and is the de facto AoT mascot. As for the other two, well, they're trolls. No other explanation for their existence is needed.

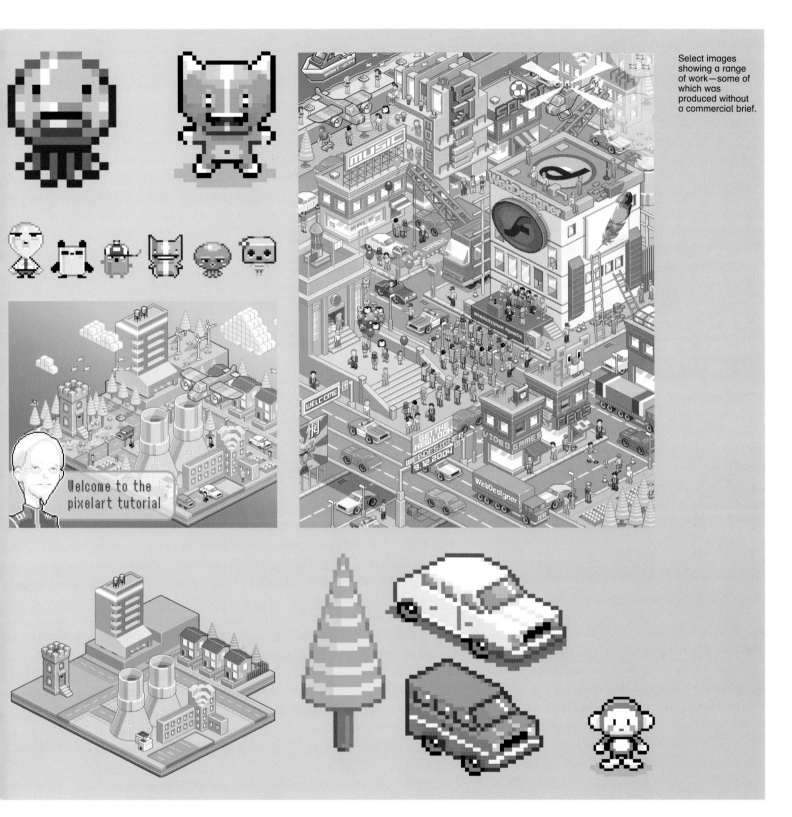

Welcome to the pixelart tutorial

Artist focus:
Army of Trolls

These images show some of the variety Army of Trolls produces. Many of these pieces were created without specific purpose, and may be used in larger works. The large, top-right image was produced for a design magazine.

Large images

Most of Army of Trolls's larger images are for clients who bring their own ideas to the table. After a bit of back and forth, a final concept is agreed upon and Gary sets to work. "I normally do a few hours' research beforehand. If it's a cityscape or a room scene, I grab as many reference pics as I can from Web sites, pictures of interesting buildings, old video games I like the style of, any item I might want to put in the picture, or anything else that gets the creative juices flowing."

"Once the research is done, I work on the base of the picture. I block out the shape of the room or the landscape then add details like roads, pillars etc. If I am working on a cityscape, I'll copy and paste buildings into the image and mess around with them in Photoshop layers until I am pleased with the layout. Then I start work on the individual buildings, normally in a separate new file, just to keep things clean. Once the buildings are finished I move them into a single layered file and shift them about until I get something I like. With isometric illustrations, I tend to chuck as much at it as I can, as I find that the more you add the better it looks."

A big A3-sized pixel illustration takes about three days for the Army of Trolls to complete, with the aid of a large library of pieces amassed from previous works.

Sprites

"If I am working on game characters, my approach is a little different. I do tend to do sketches, but I don't sit down and hand draw every frame. I sketch until I come up with a character I like, then with this in mind, I move onto the computer. Most of the time I won't bother with a scan, especially if it's a tiny 16 x 16 sprite. Once you resize your scan you won't be able to make it out anyway. So I just create my characters while looking at the sketches. If I am planning to animate them, I start with the character always face on. This means that if it's going to be used in a game and there are lots of animation routines, the transition from one to another is seamless. If I am working on a walk animation, I will concentrate on the arms and legs first, getting the walking motion looking good. Then I add small details to the character, such as the head bobbing up and down or other features."

Theory— changing hardware

This image of an arcade offers an intriguing interpretation of reality. Most of these machines actually exist in a less isometric form, and except for the open space and lack of white noise this could almost be an actual Japanese arcade.

Games

"Army of Trolls is a very small army. Of course if I am working on a game I will work alongside a programmer and sound guy, but all the artwork I do myself. Sometimes I do wish for an army of elves who would complete my work for me while I sleep."

Select images showing a range of work—some of which was produced without a commercial brief.

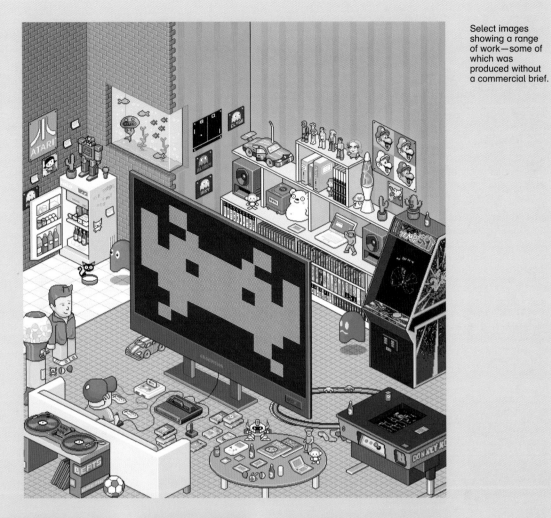

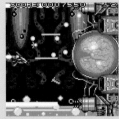

Artist focus: Army of Trolls

Most of these images are taken from game mockups. The top-right image is from the cover of an *Edge* retro issue.

Artist focus: eBoy

One group of pixel artists more than any other has moved off the screen and into the physical world of graphics and the printed page.

There's a group of pixel artists who have made it big in the real world. Three men in Berlin and one in New York call themselves eBoy, and they make pixel art that has appeared in magazines, on albums, posters, Web pages, commercials, and in other places besides. Formed in 1998, eBoy's mission is simply to give the four artists a "stage and a shared identity and a shelter from all the killers out there."

eBoy's work resembles, but does not spring directly from, video games. Only the New York quarter of the team draws from a childhood of gaming; the other three grew up in East Germany where video games weren't quite as well known. Instead they draw from other pop-culture elements, such as television, advertising, supermarkets, and Lego.

Their art runs the gamut from simple faces, animals, and rampaging beasts, to giant cityscapes filled with isometric buildings, vehicles both worldly and fantastic, designer trees, and the occasional nipple.

The appeal seems universal—even viewers without a background in gaming enjoy it. eBoy's tactics are simple. Their smaller pieces are a mix of hard edges and complicated designs rendered simplistically with oversized pixels and a very spare use of color. Larger works dish out a dual assault of this attractive simplicity with riotous complication, offering hundreds of simple, often monochromatic pieces combining to create an enormous whole.

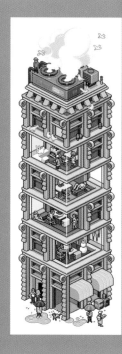

Theory—changing hardware

eBoy's work serves no real purpose; it simply exist as a kind of idealized, mesmerizing, visual cacophony. A maelstrom of vivid imagery. The above image shows many of the highlights of a trip to London.

Select images showing a range of work. The top-right image is a portrait of fashion designer Paul Smith.

Artist focus: eBoy

Two of the eBoys look over one of their riotous cityscapes. You can almost feel the intense energy.

Their art often brings immediately to mind an era of games gone by. There's a familiarity, almost like déjà-vu, about their pictures. It's the pixels and the geometric precision that cause the recall. Games like *Sim City* look almost but not exactly like the cities created by eBoy. All of us have seen icons that look, at first glance, like some of the rogues in eBoy's bizarre galleries. It's an illusion, however.

There never was an era in games that mixed the primitive graphic approach eBoy employs with the number of hues and shades they rely on. These creations spring from an age that video gaming skipped, an age of unlimited color and resolution without even a nod toward realistic use of perspective or proportion. And yet, for all the reasons that eBoy's work couldn't have come from games, they still look like they do—and that is part of their unique appeal.

There's a growing demand for the kind of art eBoy creates. Their client list includes an impressive array of global companies. Amazon, Coca-Cola, Renault, Adidas, and MTV are but four from an impressive roster. It has been remarked that eBoy might have been right at home in the 16-bit era of gaming. Their work evokes the very best qualities of the old school. They're simple, they're complex, they're colorful, and they're large. There's no doubt they might have been comfortable designing game graphics back then, but it doesn't really seem they're having out of their element now.

It's an interesting phenomenon that pixel art, created at first solely for video gaming, has moved on, beyond the games, and is now found in great numbers on home pages, magazines, and grafitti. Even unanimated, these chunky stacks of blocks are breathing with a life of their own. You can find them

everywhere, drawn by anyone, and with newer techniques and better tools, they're looking better than ever before.

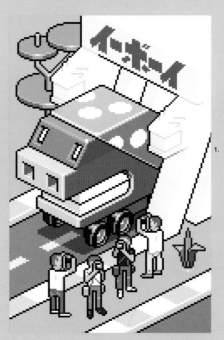

1.

These miniature images have all the hallmarks of eBoy's larger works. Chaos, excitement, and a distinct lack of empty space. Every image offers something for the attentive viewer.

Select images showing a range of work.

Artist focus:
eBoy

eBoy's works are playful, colorful, and wonderful. Part toy landscape, part video game, and part cultural statement, each image grabs your attention and holds it tight.

Artist focus: Chris Hildenbrand

Chris has an incredible range of graphics and styles, which make him an in-demand talent in the high-speed world of character design and pixel art.

Chris Hildenbrand is an incredibly prolific artist, creating graphics for one or two dozen complete games every year. He started working on pixel art with his first computer—a Commodore c64—using a joystick and Paint Magic. A few years later, like most pixel artists of the time, he moved up to the Commodore Amiga 2000. The Amiga offered proper mouse control, serious animation capability, and a 32-color palette, which allowed him to begin work on professional games. He set up his own development studio, which began releasing games for the Amiga, Atari ST, and PC.

Chris took an extended break from game graphics, moving into Web and package design in the nineties. But when Macromedia Flash was introduced and the possibility of creating and distributing games online was realized, he shifted back into game development. Since then he has been working as a freelance artist on various online, console, and cellphone games for clients such as Disney, Warner Brothers, and MTV, as well as producing his own titles with independent coders. His most successful release so far is *Heli Attack 3*, which was played over 142 million times in the first 90 days of its release.

Chris has a broad range of skills and methods that have allowed him to produce an incredible range of graphics in many styles. He works with whatever speeds his methods require, creating pixel graphics from a number of sources using various techniques.

He says, "I work best when juggling a dozen projects at a time. Creating different styles and approaches prevents me from running out of ideas or into creative blocks. I switch projects for a while and use the positive experience of one to bypass the block I face with another."

1. *Dee Jay* (GBA)

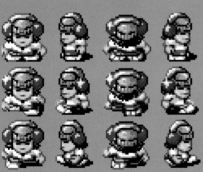

Theory— changing hardware

The game DeeJay was an amateur GameBoy Advance game. It was produced from start to finish in 24 hours as part of a competition, and included many sprites, backgrounds, and splashscreens.

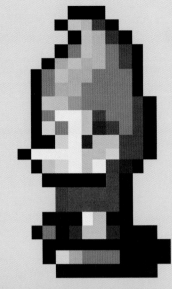

1.

TERRAN LANDMASTER

Rocket Rally

Max Rockets:
Max Turbos: 4
Mass: 2500
Top Speed: 192

EARTH'S PROUD CONTRIBUTION TO ROCKET RALLY'S LINE-UP
IS THE TERRAN LANDMASTER. IT IS MADE MOSTLY OF PURE
TITANIUM. THIS ALLOWS FOR EXCELLENT ACCELERATION AS
WELL AS GOOD CARGO ROOM. THE 6 WHEELS ENABLE
LANDMASTER TO HUG THE ROAD TIGHTLY AROUND THE
HAIRPIN TURNS OF THE MORE ADVANCED CIRCUITS, WITH THE
TRADE-OFF OF A LOWER TOP SPEED.

Roll-over a racer to read info. Click to select.

return to menu   reselect level

2.

3.

Artist focus:
Chris Hildenbrand

"When working with independent coders, the most common approach to creating the art is to play an early version of the game with placeholder graphics in order to get an idea of the game's play. Once I have a solid idea what the game is about I try to create a sample screen, working out the elements that will bring across the look and feel of the finished game to developers, managers, and producers. In many cases it means adapting to a style of graphics that suits the game and matches the target audience's expectations.

"This involved using a variety of tools—with Flash allowing the use of vector graphics and mobile platforms increasing in capability, pure pixel art is not always the only option. Using vectors as well as pixels in the creative process proves to be a helpful timesaver. Being able to scale and rotate elements without losing detail, or ending up with jagged edges and a lot of cleanup work, is the biggest advantage in the early stages of creating a character."

Chris tends to eschew the mindset that many pixel purists have. He was once banned from a pixel artists' forum because he had the audacity to create sprites without placing each pixel individually. Because of this need to create graphics from both pixel and vector sources, he avoids programs that work only with one or the other.

"My main tools are Corel Graphics Suite 12 and my Wacom Graphic tablet. It used to be a money issue: when I was working on a PC CorelDraw 3 was the best, most affordable tool. I have stayed with it since, partly out of habit, but also due to the good set tools that enable me to create my art as fast as possible. It also allows me to mix pixel art with vector graphics very easily."

1. Mangrove swamp image

1.

1.

2.

Artist focus:
Chris Hildenbrand

Chris waxes philosophical: "Using pixels as a means to create art is exciting, and groups like eBoy have come up with some amazing works. It's great to see the medium put to such an excitingly different use.

"Pixels have become a widely accepted means of illustration whether it's for games, magazines, or even video clips. It has a unique appeal that to me is a lot nicer than most of the 3D art out there, which tends to lose its character and become little more than a large number of polygons that try to imitate reality but are bound to fail.

"Most pixel art doesn't try to achieve this level of perfect realism but uses the limitations as a means to create something new and exciting. When using pixel art as a means to illustrate games, the pixels have to take on a supporting role for the gameplay. Great art by itself doesn't make for a good game, but if used properly it can engage the player and offer a lasting experience."

1. Graphical *Spot the Difference* game

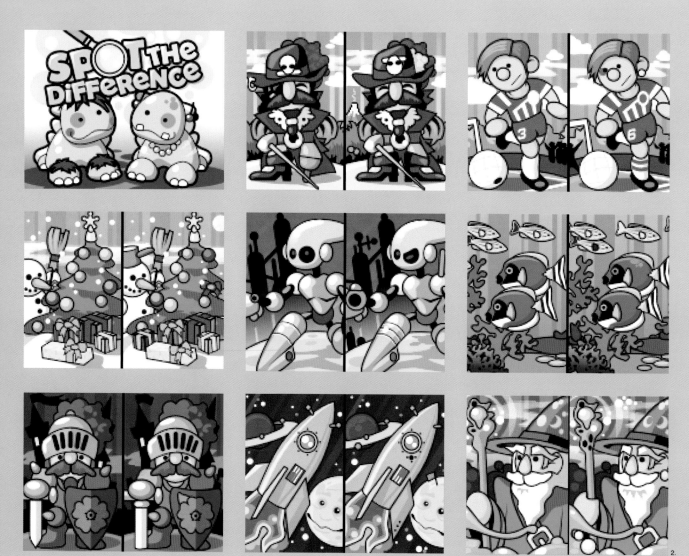

2.

Theory—
changing
hardware

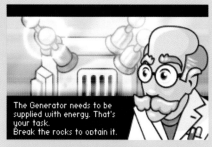

The Generator needs to be supplied with energy. That's your task.
Break the rocks to obtain it.

Adjust the rotation of the selected breaker to face the rock.

1.

1. Designs for *Gadgets Challenge*
2. Designs for *Call of Duty II*
3. Designs for *The Goobs*

2.

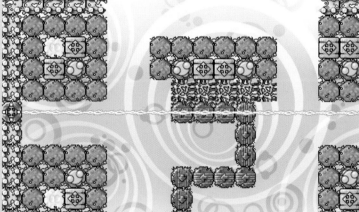

3.

Mobile developers

A huge force in the market for games is the broad array of companies that design and develop characters and games for the fast-moving cellphone world of downloads and peer-to-peer sharing of games and ringtones.

Cellular phone games are a burgeoning industry at the moment, though some suspect that the proverbial bubble may burst. Because the hardware platforms—the phones themselves—are still in their formative years technology-wise, and also because of the nature of the tiny devices, the pixel has found a new home here. Though the gameplay experience on cellphones still often leaves much to be desired, the art can be incredibly charming.

Many liken the current cellphone game environment to that of the classic console days, with smaller teams, smaller budgets, shorter development cycles, and bright-eyed ideas about potential for the future. The big difference though, is the huge number of devices these games must be ported to, the data constraints with which artists must work, and the types of games that actually sell.

Some lament that even in the cellphone world, where 2D is currently king, the polygon will push the pixel out of the limelight soon enough. But the appeal of retro gaming is strong here, and you can't do retro without the pixel! And while the games may not suit all tastes at the moment, with companies such as Square Enix, Namco, Capcom, and Hudson making significant moves into the mobile arena, it's only a matter of time before a compelling "killer app" shows up on the radar.

On the following pages, we'll hear some words from developers on the subjects of pixel art creation, the cellphone industry, the tools they use, and where things may be headed from here.

"The cellphone is a unique platform for developers. Games are neither released on one or two platforms, as in the case of console,

1. Nokia N91 cellphone
2. Nokia N80 cellphone

delivered for a minimum and maximum hardware spec, as in the case of the PC. The big difference is that a cellphone game has to look good and play well on potentially hundreds of phones, each with its own unique resolution, memory, and processor limitations.

"Perhaps the biggest art challenge is getting the same game to look great on almost every phone. With hundreds of handsets, the graphics have to be optimized into families of phones that have similar screen resolutions. Some games use the screen space in exacting ways and so they require more careful attention to element placement. Additional limitations and quirks of some handsets can require some fairly radical alteration or redesign of the art altogether to try to approximate the same gameplay experience as a higher-end handset."

Charles Barnard
Art director, Glu Mobile

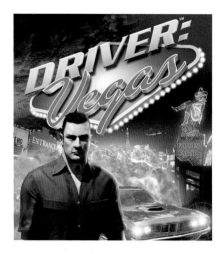

1. *Driver*,
   Glu Mobile
2. *Jamaican Bobsled*,
   Glu Mobile

Mobile
developers

"Games should be short enough for consumers to play them during a brief period of free time, such as between classes or meetings or while waiting in line. There may well be new ways to do this, which if realized, could expand the games market exponentially. Whether a game world lives on while the user is not playing—the gamer is alerted when a matter needs attention or a quick decision must be made—or the application provides constant game persistence or a faster startup time, getting the user into the action should be the goal.

"The future of cellphone games will begin to consider the reality of the phone experience and factor it into the games that are created. Instead of cloning games and gaming experiences from other platforms and devices, games will begin to address the way people interact with their device, the time people spend using a cellphone game, the frequency of use, and the connected nature of phones that allows people to collaborate and compete. "

Cris Cook
Lead designer, UIEvolution

"The most frustrating thing about producing art for cellphones is the file- size restrictions. Screen resolution is not such a problem; it's actually quite appropriate for pixel art. But it's sometimes frustrating that you need to limit the amount of pixel art you deliver, because of memory size— but then again, you only really use lots of space when you screw up!

"The greatest opportunity, then, is that everyone carries a gaming platform with them everywhere. The nature of the phone dictates what games are suitable on them and right now the industry is in a learning phase. I think that when the developers, publishers, and consumers get more accustomed to the format it will outshine all portable devices in gaming."

Henrik Pettersson
Artist, Jadestone

1. *Kodo* screenshots, Jadestone

1.

Mobile
developers

1. *Spirits* screenshots, Jadestone

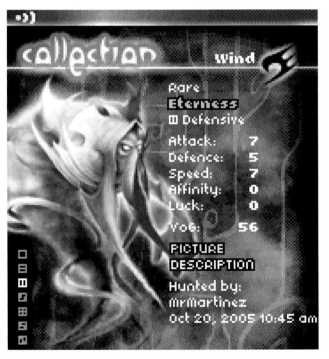

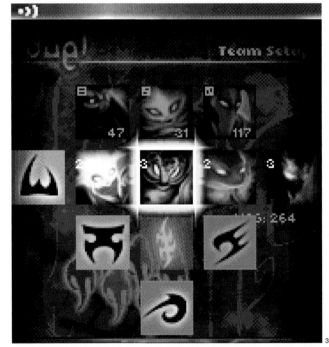

"The [pixel art] process is the same as in most visual art. You need to start with a sketch. Generally I make a high-res sketch, larger than I need, and scale it down. From there I tend to go in different directions depending on what I'm doing. You have your outlines, your blocking in of color, form, highlight, and so on. It's a question of tools, restrictions, and requirements as with other media.

"What tends to happen is that you get new names for old concepts—pixel art isn't that special or different. It was the same when I did a short course in tattooing. It's also a fringe section of visual art, but when you get into it you find that it has the same issues to wrestle with, it is just the tools that are new. That said, I will readily admit that reading every pixel art article that I could find really improved my work, and honestly I'm looking forward to my own further improvement.

"I don't consider myself a hardcore pixel artist so I do use Photoshop, which is my preferred software whether I'm doing pixel art or high-res art. I guess when the day comes that I need to do more advanced animation, I'll have to find something else. I do see why many purists say that you shouldn't use Photoshop though. It is often tempting to "cheat" and let the software do the dithering and such. But hey, whatever gets the job done is fine by me."

**Henrik Pettersson
Artist, Jadestone**

"The tools used to create and edit images for cellphones are more specialized than those used for the PC or console. For one thing, we rely on lower bit depth and exacting color palette control to generate images that will fit within the limited memory of the phones. This requires tools and software that are not too commonly used in PC or console development today. Admittedly, some of the best tools seem to come from the past. One of our artists at Glu does most of his work in a program called D-Paint, which is almost 15 years old, and provides him with exactly the level of control required. Other artists use tools like Pro Motion, which is a pixel art and animation tool designed with the handheld market in mind.

"Developing the art itself is not unlike developing graphics for PC games 15 years ago or on some of today's handhelds, with their smaller memory and slower processors. The good thing for cellphones is that we have the legacy of those old PC graphics to draw on, and our artists are well versed in those techniques, so they can maximize the graphic quality and content."

**Charles Barnard
Art director, Glu Mobile**

1. New game for Jadestone, Glu Mobile

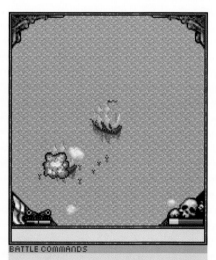

1.

Mobile developers

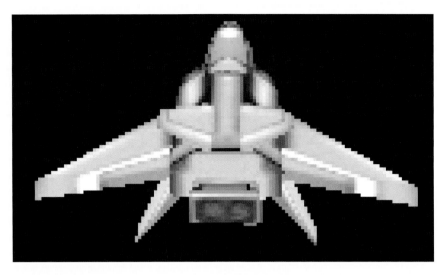

1.

"Being a retro gamer myself who is intrigued by the challenges of creating good-looking graphics on a limited platform, I was attracted to working on cellphone gaming with Glu.

"Though our games do not target the retro gamer specifically, many of the people working on the products here are retro gaming fans and frequently look for opportunities to pay homage to the gaming classics of yesteryear. Retro titles like *PacMan* and *Galaga* are big sellers on cellphones, and it's a good opportunity for the companies that hold these older titles to revive their franchises and present them to a fresh market."

**Charles Barnard
Art director, Glu Mobile**

"Retro is popular in cellphones precisely because it's retro. It's nostalgic. Just like the smell of apple pie might remind you of your grandma's house, or a classic Eighties' tune will take you back to your younger days, pixel work will remind gamers of when they first started to play games and which ones they fell in love with. I remember playing *Wonder Boy in Monster Land* at the laundromat while my Mom washed our clothes, and it makes me happy to think of those times. Younger gamers won't remember those early games but it's been implanted in the gaming culture at large. Games like *Duck Hunt*, *Super Mario Bros.*, *PacMan*, and *Space Invaders* have become icons in today's world and can be recognized by countless individuals whether they play games or not.

"Of course it's always good to give glimpses of past games in current designs, and most artists and designers want to do just that. It pays homage to games that influenced them. Obviously we can't go overboard with it, but taking reference from games gone by is a great way to learn from the past and to subtly remind people about where games come from. So what more can we do with pixel art? Well, it's going to take some forward thinking. Look at Nippon Ichi, they've gone and made 2D/3D hybrid games like *Disgaea* and proved that such an art design can work commercially. Arc System Works makes huge sprites for its *Guilty Gear* series, which then begs the question: why not make more high-resolution 2D games? Without question the systems of today can handle this notion. But then as pixel work gets to be a higher resolution, it changes from pixels to illustration. But that's what technical advancement is, and evolution is inevitable. Pixel work will evolve to a more illustrated format in that sense. But pixels will never go away because of the nostalgia factor. People still want to see pixels and so we'll be able to play with them for a while yet."

**Randy Van Der Vlag
Lead 2D artist, Gameloft**

1. *Fruit Twist* mockup, GameLoft (Cellphone)

2 Gangster character study, GameLoft

1.

2.

Mobile developers

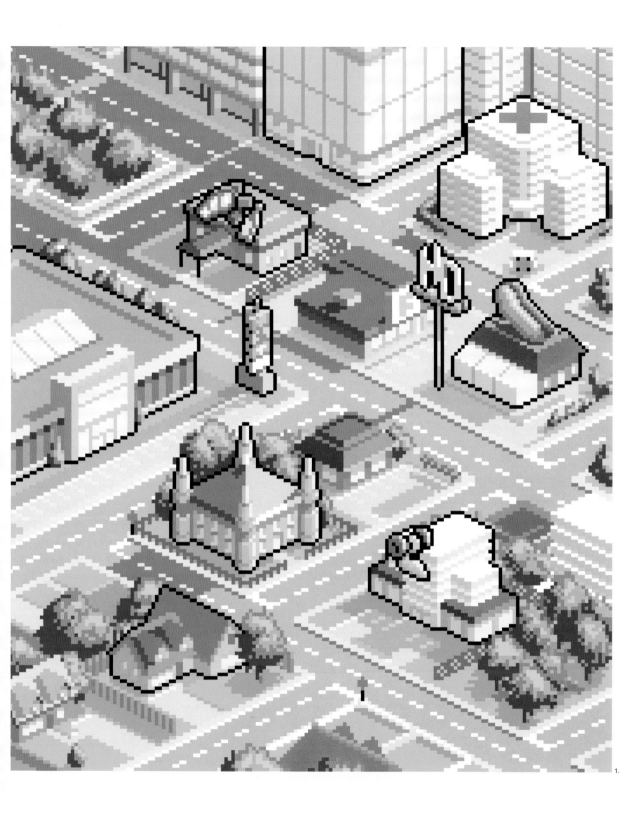

1.

Mobile developers

"There will hopefully always be a place for a simpler style of art on cellphones due to the small screen resolution and limited palettes. This holds a value all its own. There is no reason to believe that resolution, palettes, and memory will continue to be restricted as much as they are now, given the rate of technological advancements.

"If you're looking for a lifelong career within visual arts, I wouldn't recommend pixel art as your only skill. At the same time, I think even in five years' time, there will still be people carrying cellphones with less than perfect 3D performance, so there should still be some room for the pixel. In 10 years' time, who knows? Hopefully, someone will still make games that look like the old 2D games, even when we can no longer make pixel-perfect images.

"I don't find that cellphone art restricts what I can do—the platform itself can't be considered restrictive,

as pixel art actually thrives on restrictions. That is what drives all innovation within pixel art. Those restrictions will most likely disappear sooner or later and the curse of 3D will take over, if history is anything to go by."

Henrik Pettersson
Artist, Jadestone

"In terms of making cellphone art development easier, making the phones more consistent in memory, graphic processing ability, and resolution would help a lot. In other words, the industry should develop consistent standards that would span across more phones.

"This would give art production very clear guidelines for a broader range of phones and ultimately reduce production time, as we wouldn't have to create art or code for as many special cases. This would also mean that any new phones coming out would be a bit more

predictable in the way they would handle graphics. I realize that isn't realistic for such a rapidly expanding technology, nor would it be preferable from a marketing perspective, but it would sure help the artists!"

Charles Barnard
Art director, Glu Mobile

"I think the future of pixel work is on the Internet more than the cellphones, and it is establishing itself as a proper artform. There are lots of pixel art sites and pixel communities that one can go to and learn how to push pixels. Pixels have gone past just being in games and have since extended to become a genuine form of artistic expression. And there isn't any reason that it shouldn't be this way, because what is pixel work if not a modern version of pointillism or mosaic tile work?

1. Military game
mockup screens,
GameLoft
(cellphone)

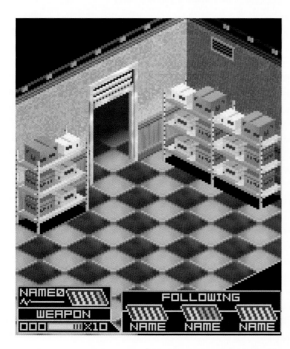

1.

Mobile developers

"I think that independent games are the future of the pixel. Small teams have already begun to create fantastic pixel-based games like *Cave Story*, *Mimi Pan*, and *Tag Der Arbeit*. There is a drive and a desire for pixel games, and the Internet will be the place to find the most creative among them, because these games will be made by people that love pixel work and that don't have to risk anything when making them."

Randy Van Der Vlag
Lead 2D artist, Gameloft

Blue Label Games is a small cellphone porting house based in southern California. Given the sheer number of phones they need to port to, they're in a good position to talk about the nuts and bolts of getting pixel art into phones. The tips and tricks overleaf are written by Paul Zirkle, Core Engine Programmer for Blue Label Games.

1. Skiping Stone,
   I Play (cellphone)

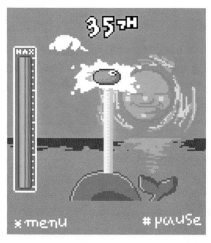

1.

Cellphone tips and tricks

Paul Zirkle, Core Engine Programmer for Blue Label Games, gives us his hints and tips for designing for the cellphone.

Part 1: Sprite formats

When creating art assets for a 2D game, one of the types of assets artists create is character art—specifically, animation sprites. The exact formats of where animation frames are located in an image (or across multiple images) are dictated by the programmer, so I can never give exact information. However, two common sprite formats are "Film Strip" and "Jumbled-Up Piece of Crap" (JUPOC).

Film Strip is very simple: each animation has the same height and width. It's easy for the programmer to use because getting to the particular part of the image with the right animation is just a simple mathematical function: (animationNumber * animationFrameWidth).
The problem with this format is that it can create lots of unused space. Consider a character animation set that includes a person who is walking (30 pixels high by 10 pixels wide average), but also a sequence where they are laying down dead (30 pixels wide and 10 pixels high). Because each animation has to have the same height and width, you will end up with giant amounts of unused space—each frame has to be 30 pixels by 30 pixels.

The answer to this can be to break up animation sets into similar-sized frames only. So in the case above you'd have to have one image for walking animation, and a separate image for the death sequence. Unfortunately, this requires more logic for the programmer since he now has to write logic for deciding which image to use, but this may be acceptable.

In some cases however, there is still too much unused space. Jumbled-Up Piece of Crap (JUPOC) resembles 3D model textured skins—the kind generated in Alias's 3D software MAYA.

1. Trane, from Marc Ecko's *Getting Up*, Glu Mobile

1.

It basically relies on the exact knowledge of how big each piece of a sprite is and crams it all together into one giant image with little to no unused space. However, this requires that a look-up table be stored somewhere that can tell the programmer where in the image a particular rectangle for a particular sprite is located, and how much that rectangle is offset from the average frame. This obviously takes greater coordination between programmer and artist, and is more work for both.

1.

File formats and memory usage

When dealing with cellphones, we usually need BMP or PNG format files. Some phones support only BMP, and those that support PNG usually have issues with different encodings, so they can be tricky in their own right.

On cellphones, we have to worry a lot about memory. The word "memory" is bandied about quite a bit though, and it can be confusing because programmers are concerned about two kinds of memory: one is File space, and one is Heap space (aka RAM or system memory). The first one is obvious— how much space the image takes up on the file system of the phone. The other is how much memory your image will take up in RAM when loaded and ready to be displayed on the phone. So here's how we figure it out. If you save a 100 x 100 BMP image in 8-bit color

depth it will take up less file space than the same image with 16-bit color depth. However, on a cellphone there is only one supported native format. When that image is loaded and ready to be drawn to the screen, it must be converted to that format. Many phones are now 16-bit, but some are still 8-bit, and there are even strange formats like 10-bit (i.e., 3R4G3B).

A programmer may ask you, the artist, to be conscious of how many colors you use to create assets, and to save them in indexed color mode to save file space. However, that same programmer may tell you that only so many images can be shown on the screen at the same time, since even if it's saved in small format, that image may blow up to the size of a full 16-bit color image when drawn to the screen. There is a programming technique called JIT

(Just in Time) in which you load an image into memory just long enough to draw to the screen, then release it again, which allows you to play with more images than you have available memory. However, this technique makes the program run slower due to the excessive loading times between drawing frames. JIT is a four-letter word.

Another consideration in memory usage is the stuffing of files with multiple images. Essentially, every single image has a color table, but many images use the same color table. If you were to put the two images together in one image file, then only one color table would be stored. However, the programmer would also have to add logic to his program to be able to draw only the appropriate portion of the image file to get the right image out. Not only that, but the entire image file would have to be loaded into RAM

1. *Dominoes* game,
Blue Label
Games
(Cellphone)

1.

Mobile developers

when only one image was to be displayed. That is why multiple graphics are typically put into the same image file only if they are likely to all be needed in memory at the same time. For instance, consider the HUD of a driving game. You may have an image for the dashboard and one for the speedometer needle and another for what gear you're in, etc. Another application of this idea is character animations or sprites.

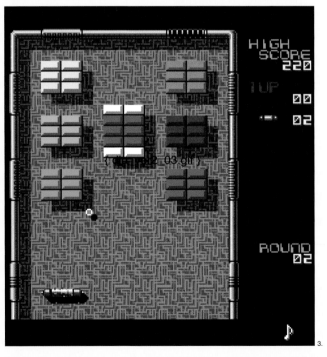

1. *Bubble Bobble*, Taito (cellphone)
2. *Wrestling* game, Blue Label Games (cellphone)
3. *Arkanoid*, Taito (cellphone)

Cellphone tricks
and tips

Resolution dependence and independence

When making a cellphone game, the goal is often to put it on as many different handsets as possible. This means it will run on devices with different screen sizes and different amounts of file and Heap space.

Some game designs are more suited to this type of thing than others. For instance, consider a game that scrolls in all directions (up, down, left, and right). In such an instance, if the screen size of the first "target" device is 100 x 100, and you're trying to go to a phone that is 80 x 80, the part of the game that scrolls doesn't have to be changed! The player simply gets to see less of the play area and will have to scroll more.

However, consider a game screen that requires a graphic that exactly fits the size of the entire screen. For example, let's say the designer wants the main menu to have an awesome-looking background and title image. If an artist ends up creating a 100 x 100 image for that first phone, he or she will have to create a separate 80 x 80 image for the next phone (or the programmer will have to take the image you made and resize it, but it'll probably look much worse that way).

Now consider a menu background where the artist creates two different images. One is the title graphic, which is made only 60 x 60. The other is a tiled background image (let's say 20 x 20, but it doesn't matter because it's tiled) that will

fill up the entire screen behind the title graphic. Now the menu is ready to work on phones for a flat color fill provided by the programmer and just make sure the title image has the same color on its borders.

Even if the game scrolls, and the art design is intelligent enough to handle small changes in resolution, artists may still have to create groups of assets for major groups of resolutions. Here we have three categories: Large (176 x 204), Medium (128 x 146), and Miscellaneous Small. The producers of newer phones seem to have picked up on the idea that creating phones with the same resolutions will aid in porting applications, which means more applications for their phones and therefore higher sales.

1. Game for Jadestone, Glu Mobile

Mobile developers   Henrik Pettersson is responsible for the art in this new Jadestone-developed, Nokia-published mobile title.

Older phones (with chronologically smaller screen sizes) are totally different from one another—hence the "miscellaneous" title, with no average resolution given.

Another tactic in the battle for a more efficient art pipeline is the use of vector and/or 3D graphics. Because it is assumed that artists will be creating specific resolution renders from their original art, it's very easy for them to create a set of Large, Medium, and Small animations from their 3D modeling program. Similarly, with vector graphics, they can create a title graphic that can be sized to any resolution and maintain a crisp appearance. The problem with these is the amount of overhead it takes to create the original art asset.

When you realize how many different sets you'll have to create from that work, however, you will realize how much time you are saving yourself in the end.

1. Upcoming game, Jadestone, Glu Mobile

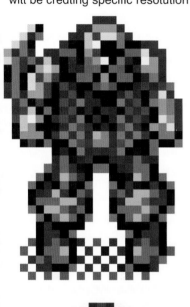

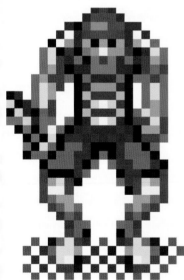

1.

Production Cycle

As mentioned, a cellphone game should be made for as many handsets as possible. Currently, one development house will create an application for one or two phones, and then the publishing company that hired the developers will take the code and art assets and give them to a separate porting house to put onto handsets they were never intended for (and ask for new game features at the same time).

As an artist, you might be involved in the whole cycle or just the initial part. You may first contract with the developer and then get contacted by the publisher to work with the porting house afterward. Your contract with the initial developer might state the requirement for the original assets to be submitted and passed on to the porting house for their own use. If you're a part of the design and development team, you may have to do a lot of remakes and edits, but this can be typical of all art design jobs.

It's helpful for us as a porting house to have the originals, as many of our programmers do touch-up art and make minor changes as prescribed by the clients—particularly when their legal department comes in and says one of the character's hats looks too much like the symbol for the University of Texas and we have to change it, after being resized to 8 x 8 pixels, or when the ninja star icon from a certain classic remake looks too much like a swastika. There usually isn't much communication between the developers and the porting companies after that, so the artist also usually gets beyond the reach of the porting houses. Keep in mind, though, programmers are not artists. We may be able to modify some images here and there, but sometimes we have to call in an artist to do the job right. Furthermore, managers don't like to see their expensive programmers getting paid for work they could convince an artist to do at half the price.

One last piece of advice for aspiring cellphone pixel artists: start big, work down—graphics are harder to size up than down. Just because the current contract says the application is for phones of X, Y, and Z size, don't assume the developer won't come back later and ask for A, B, and C as well.

1-2. *Monkey on Your Back*, Capybara (cellphone)

1.

2.

1. *Monkey on Your Back*, Capybara (cellphone)

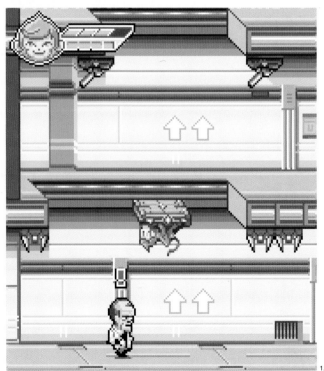

1.

Cellphone tricks
and tips

Results

Most of our art is handled through contract work. Our porting house employs only about 10 programmers, and four to eight manager/QA/ miscellaneous characters, and none of them are full-time artists of any kind. I've found most of the art guys we've worked with were new to the deal, and I've had to explain what we want to enough of them to be able to repeat it here. This means that some cellphone companies are willing to accept relationships with newer artists that may not be familiar with the setting, as long as they have promising portfolios.

Of particular interest with our *Dominoes* game is the opponent imagery. This game only targeted medium to small devices (the large sizes weren't even out when this was first being created), and so there is only one size for every image. The game screens are coded in such a way that the image is centered for smaller screens, so that only the important part of the image shows up. This allowed us to use large-sized images even on the smaller handsets, at least theoretically.

Eventually we did come across a handset that didn't have enough memory, and then we simply had to size the images down. But in most cases we could just crop the large

ones instead of resizing them. For *Frank Shamrock Wrestling*, we have a few animation filmstrips. The close-up animations are broken into multiple images because only one of the animations would be shown during that close-up screen (so we didn't need to load the others into Heap memory at that time, because they wouldn't be used). The characters in were CG, rendered to file per requested resolution.

Paul Zirkle
Core engine programmer
Blue Label Games

1. *Dominoes* game, Blue Label Games (cellphone)
2. *Wrestling* game, Blue Label Games (cellphone)

1.

2.

3.

Mobile developers

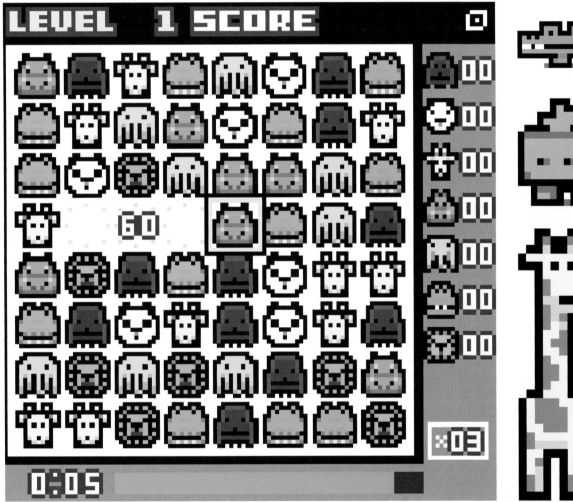

1.

Cellphone tricks
and tips

Cabybara's creation process

A step-by-step guide to the creation of a sprite — or, "How I Learned to Love the Zoom."

Words by Nathan Vella, art by Nathan Vella and Anthony Chan, Capybara Games Art Team.

The first step in the creation process is to sit down with the key creative members of the team and really figure out what is expected. In mobile, it is extremely important to gauge expectations and set realistic artistic goals for a game before getting started. Platform and carrier-set file size limitations, combined with the lack of hardware flipping, scaling, and rotation can cause problems in any game. The goal of an artist is always to make the game look as good as possible while managing the limitations of the device. In the case of mobile and 2D art, this limitation is primarily on the file size.

When discussing game art, in this case character creation, start discussing goals very widely, then narrow them down as you go. The first thing we figure out is very straightforward: is the game meant to be more realistic-looking, or should it look more cartoon-like or stylized? In the case of license-based games, the material often calls for a more realistic style. Other games allow the pixel artist a bit more stylistic freedom, and allow them to expand their color choices, character proportions, and add more of their own personal style.

Another key high-level decision is the view or perspective the game will take. This should be gleaned from the game's design document, but definitely warrants discussion as well. If it's an adventure game or an RPG, will the look be top-down? For example, is it forced-perspective, like *Final Fantasy* and *Legend of Zelda*; or isometric, like *Popolocrois* or *Mario* RPG? If it's a platformer, will you be sticking with a straight-on profile, like *Super Mario* or *MegaMan*, or will the

profile have some forced-perspective, like *Donkey Kong*, or as seen in this book, Capybara's *Monkey on Your Back*? Every style of game has at least a couple of different views or perspectives that can be used, each with advantages and disadvantages. This decision will have a direct affect on the creation of the character, both in terms of the artistic approach, and how you manage your file size.

The next step is to decide the more specific aspects of the art direction, character species, gender, outfit, weapons/tools, and so on. Again, much of this is dictated by the game's design and should show up in the design document.

Once the creative team has had a chance to solidify the broad strokes, we begin concepting. For this example, our game calls for a rock 'golem' character for a *Zelda*-style adventure game, with an overhead forced-perspective view. An important step at this stage is to work with the creative team to find references. Much like how directors, when shooting a specific type of film, will screen the classics of the genre, so too should pixel artists make references to the classics of their genres. From these references we see what was done right, and learn from it. Pixel art is rather old compared to other styles of game art, so there are ample references from which to draw.

Concepting here at Capybara is pretty wide open. Sometimes artists choose to concept via pen and paper, digital illustrations, or they concept in pixels directly. Each option has its advantages, but really it comes down to whatever you feel will work best to illustrate where you want to take the game artistically. In many cases, we choose to do illustrations first because it is a very good way to capture the intended feel of the game. Illustrations also allow the pixel artist to have a strong base to work from, and can help keep the art moving in a consistent direction. In this case Capybara artist Anthony Chan did a black and white concept illustration of our Rock Golem in Photoshop using the Brush tool (figure 1).

The first step is taking the rough concept art and adding some color to it with the Brush tool. I work in bigger blocks of color, for reasons defined in our next step. Adding color is especially important because the color really helps to clarify the light source. Having one distinguishable light source is a key element of pixel art, because even in such a small size it helps to define the shape of your figure. This light source should remain constant throughout all the art in the game (figure 2).

Now that I have added some rough color to Anthony's concept, my next step is to shrink it down to a much smaller size. This helps me to understand which of the character's nuances will translate into pixel form. In the previous step I painted in larger blocks of simple colors, with one defined light source. When shrinking down art, even this drastically, the blocks of color and light source are maintained, and will ease the process of translating this into a sprite. I personally tend to use Photoshop (CS versions) Bicubic Sharper mode of image resampling when shrinking raster artwork. Even though this creates a lot of anti-aliasing and blurs the image, it allows for the general form of the illustration to be maintained (figure 3).

For colors, I use Photoshop's Eye Dropper tool and grab the colors I feel are the most important, running the gamut from light to dark. For each color I paint a small patch of a palette so I can easily access exact colors, ensuring that during final optimization I won't have 10 slightly different variations of one color. One way to deal with the mobile format's file size limitations is to work in a very limited color palette, so for this character, I am going to grab 13 colors from the concept, define two colors for shadow, and leave one open color slot for transparency. This makes the total colors 16, which should be more than adequate. Later on during optimization I may choose to crunch the colors further. (figure 4).

1. Rock Golem black & white illustration
2. Rock Golem color illustration
3. Rock Golem shrunk illustration
4. Rock Golem color palette

1.

2.

3.

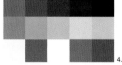

4.

Cabybara creation process

Now that I have an image to work from via the shrunken illustration, and I have my colors selected, I can start "pixelizing." The first thing I do is to get the general form of the character down. I'm not trying to copy the concept illustration; rather, I'm trying to communicate the most important aspects. I use a dark color and outline its torso, head, arms, and legs. This is meant to define the form of the character roughly—the actual definition will come later when we start adding color. When outlining the shape, it is very important to pay close attention to your lines: pixel art is all about attention to detail, and wrongly positioned pixels, even just one, may change the character from being rounded to jagged. I make sure my outlines form consistent lines so I don't end up with an arm or head that has an angle that looks wrong in pixels. Working zoomed in to 1,200–1,600 percent, then zooming out to 100 percent will

make it very clear which lines work, and which appear too jagged or unclean (figure 1).

Once I'm happy with my rough outline, I move into rough coloring. I have my color-blocks on my canvas, and from there I pick what I want to be my base color. In my case I chose a midtone and fill in the outline. It is a personal preference of mine to add light and dark afterward, some people may prefer to start dark and go lighter, or vice-versa. After the mid-tone, I'm next going to add larger chunks of shadow, based off the shrunken version of the illustration we have. With the consistent light source, dark shadows are cast between the legs, between the arms and torso, and on the far side of the torso. Since I am going for a less serious look, I won't avoid using darker outlines. With the rough shadows in place, I grab my highlight color, an off-white, and lay down

some rough highlights based on the light source. The sprite is still lacking features, like face, hands, and other details, but those will come soon. It may seem extremely rough, but think of it as working on an underpainting or laying a foundation upon which you will build (figure 2).

At this point, the sprite still lacks a lot of definition. To increase the definition, I'll start adding variations of shadow, midtone, and highlight colors. The amount of variation and placement will really define the feel of the sprite—lots of variation will create a smooth, more 3D look, while less variation and more blockiness will keep the sprite cartoony. In this case, the lack of variation at points creates the rocky texture of the golem. At the same time, I will be adjusting the actual form of the sprite, adding rough hands, a rough face, defining parts of the torso, and so on (figure 3).

1. Rock Golem outline, zoomed in and out
2. Rock Golem, early shading stage one

1.

2.

Mobile developers

This step revolves around basic trial and error, and is about finding the happy median between a smooth and cohesive-looking sprite, and the style you desire. The history tab is your friend, and using Photoshop's Snapshot function in the history tab at key junctures will help you if you make a massive change that you end up disliking. At this stage, I am constantly zooming in and out from 1,600 percent magnification where I apply the color, to 100–200 percent where I gauge the effect of the color placement. Using Photoshop's hotkey combination of "Control +" and "Control –" will keep you zoomed at equal ratios and make sure you don't end up at an odd percentage of zoom that will skew your pixels. For this stage of coloring, I work in cycles—add color, zoom out to measure the impact, zoom in, adjust, and repeat. Since we are dealing with pixel art, a single pixel can make an enormous

difference. Pixel art is very impressionistic—a single pixel can equal an eye (figure 4).

Now we are working steadily toward a finished sprite, and only the addition of fine details remains, and this is actually the brunt of a pixel artist's work. I add and remove a few pixels, keeping with the perspective of the character to represent the eyes, mouth, fingers, and so on. In many cases, small groups of pixels, or even single pixels do the job. Another key aspect is to start eliminating the uniform outline and replace it with more variation to match the light source and really accent key lines to create some separation between parts of the sprite.

3. Rock Golem
   early shading
   stage 2
4. Rock Golem
   middle shading

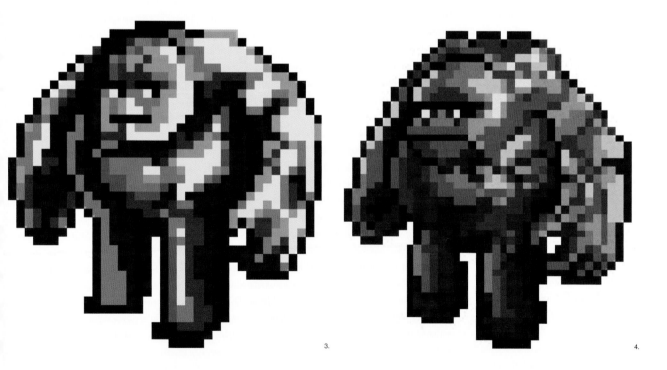

3.

4.

Cabybara creation
process

I also want to make the sprite look a little more textured, so I add some jagged rock protrusions as well a few pixels at a time. I take liberties to move the legs and arms into slightly more dynamic and "pixel-friendly" positions. Again, this stage requires the constant cycle of adding pixels, zooming out to gauge their effect, and zooming back in to adjust a pixel here and there. Once I am pleased with the details I have added, I zoom out and survey the sprite, making sure I kept to the constant light source and perspective (figure 5).

Fine tuning will help clean up the stray pixels that don't properly represent the angle or lighting. As usual, I cycle from zoomed in to zoomed out to survey the changes. A final touch comes through adding a shadow, again trying to keep fairly consistent with the light source, though in the case of the character's shadow, complete accuracy is not necessary—it's better to make a clean-looking shadow than to be completely accurate. My shadow is very round and simple, as this helps keep the viewer's attention on the character while still placing

him in the game world. Complicated shadows don't come across very well in mobile, and can end up being rather distracting. (figure 6).

Now that the sprite is done, we use Photoshop's Save for Web feature. In this pop-up window, I select PNG-8 as my file format. Next, I look at my color table to see the colors I used. Clicking on the option arrow lets me choose how my colors are sorted—I choose "By Popularity" in order to arrange the colors from most-used to least.

From there, I can choose to eliminate single colors by dragging the undesired color into the trash bin. In this case, I am happy sticking with 16 colors, so no trashing or dithering is needed. I hit save, and we have a finished sprite.

5. Rock Golem, end turning

5.

Cabybara creation process

6.

Cabybara creation
process

Sample character portrait creation

Words and art by Chris Hildenbrand.

The creation of character art for the dialog screens of the GBA/DS title *Milo's Quest: Dragonpearls* is a good example of vector use for pixel art.

2.

1 In order to find some inspiration and create a look that is natural, I started out with a search on the Internet for a young male face for the games hero.

2 The sketch was brought into CorelDraw as a bottom layer and vector shapes were created on top in order to give the illustration a very clean inked look. It also allows quick alterations, eg. the eyes were enlarged to give the character a younger, more boyish look.

3 Color and basic shading were added still using vector shapes for easy editing.

4 The vector shape is then copied into CorelPhotoPaint as a new object and scaled down to size.The edges were manually cleaned up to allow 1-bit transparency and give the portrait enough contrast. Magenta is my personal favorite for the transparent color as it is rarely used in my work while standing out enough to see stray pixels.

5 The portrait shape was masked and new layers added on top. Highlights and shadows were painted on using an airbrush tool and different layers with different transparency settings.

Sample character
portrait creation

105

1. Stage 1
2. Stage 2
3. Slime

1.

2.

Sample character creation

The task at hand was to create a detailed medium-sized enemy for a role-playing game. The "Snow Wolf" had to stay within a size of 200px x 160px with no color restrictions.

Snow Wolf creation stage one

I started out with a blank canvas in Corel PhotoPaint set to double the final size. Working digitally from the start using a graphic tablet, makes tasks like the inking a lot easier—as I love my Undo button. With a pencil tool and a light gray color, the main shapes are laid out and later outlined with a black pen tool. This sketch was then reduced to the final size. After enhancing the contrast and altering the tone-curve, I reduced the sketch to two colors (black and white) and manually

cleaned up the outlines. Coloring the main shapes with the Fill tool allows a quick color composition. Afterward the black outline was made into a separate layer with additional layers for highlights and shadows added above.

Snow Wolf creation stage two

Setting the colored areas as a mask allowed easy shading without covering the background. Keeping the background clean is essential to allow transparency once the art is displayed inside the game. Highlights and shadows were added using an anti-aliased pen with varying degrees of softness on separate layers.

1.

2.

1. Troll
2. Crab
3. Snow Wolf

4.3:
Sample character
portrait creation

107

Sample character optimization
Depending on the game, its engine, and the use of the resources, it sometimes makes sense to break up a character into its elements.

For the Flash game *Heli Attack 2* the coder used this technique to allow the players guns to point to the mouse's position for aiming.

The sprites were rather simple as was the game. The whole artwork took just over one day but the game became a huge hit online—winning miniclip.com Game of the Year in 2003—and sparking the demand for a sequel.

For *Heli Attack 3* the player movement had to be more elaborate allowing the use of over 30 different weapons, as well running, crouching, jumping, and climbing. Being an online Flash game restricted the use of graphics in order to keep the file-size within limits.

The new look also had to be more futuristic yet in line with some of today's weapons the player would be using in the game.

As a 2D, side-scrolling game *Heli Attack 3* artwork only needed to be done in one direction and then flipped in Flash.

Even though Flash is able to rotate bitmap graphics for the player sprite, the main body was tilted in several images to ensure a decent display quality of the facial features.

1. *Heli Attack 3* sprites

1.

2.

Mobile developers

The legs were animated separately using an eight-frame cycle. The standing position was used as a base and the animations were painted in layers above it. With CorelPhotoPaint's animations feature the run-cycle was finetuned and then exported to 1-bit transparent PNGs.

On top of those two elements the arms and weapons are displayed with animations for reloading and empty shells. The rotation to point them at the mouse's position is done in the game.

While most of the game elements were "pixeled," the interface images were created using vectors to determine the look and coloring of the health indicator in the game.

1. *Heli Attack 3* sprites

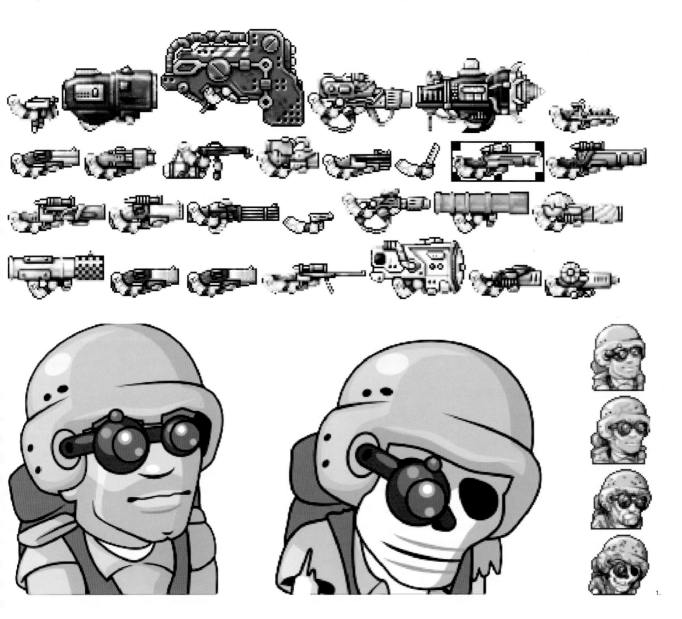

1.

Riva Jan Halfer

Whether he's pixeling for cellphones or PCs, here is one artist who has a unique and imaginative vision.

Jan Halfar's colorful art brings to mind a Victorian, steam-driven world of clockwork birds, floating castles, and industrial malevolence. A handful of small images give a glimpse into an entire world, similar to, but separate from, ours.

"When I started with pixel art years ago, I would create images from start to finish using only the mouse and a paint program. Lots of these early attempts just didn't come out right, mostly because of questionable composition or character proportions that were somehow wrong. This changed when I picked up an old hand scanner, and began scanning sketched and inked images as a basis for my pixel creations. I still use this technique for larger works.

"With a scanned version of the sketch as a guide, I start by filling it with solid, basic colors. This helps me define the basic shapes of the components in the image. It's very easy, at this stage, to tune the colors, manipulate different parts of the foreground and background before losing myself in the tiny, pixel-by-pixel details.

"If a new pixel artist were to ask my advice, I would tell him or her: Dust off what you know from your classic painting classes. If you never attended one, do so. Back in my high school art class I learned the basics of composition, lighting, and color, and these fundamentals apply in all kinds of art, including pixeling.

Next, don't believe that some pixel technique tutorial will miraculously improve your skills just by reading it (as many newbie artists think today). You have to pixel (or paint) images yourself, lots of them, to get better. Also instead of tutorials, I would recommend taking some masterpieces from true pros in pixel

Mobile developers
Small header

These images show the progression from pencil sketch to fully realized art for an old-school mobile shooter Halfar is working on with his old friends, Digital Monks.

art, magnifying them, and dissecting them pixel by pixel. Examine the techniques used by each artist and study how they dealt with various problems. You might recognize a solution to a problem you're facing and then try to reproduce it in your own art. That's how I learned and continue to learn more and more about my own art."

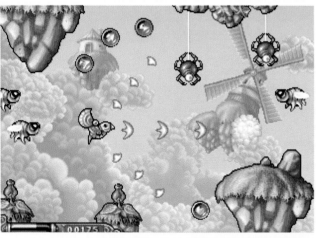

Riva Jan Halfer
Small header

Clockwise from top right:
A skeleton warrior design
for an unreleased
dungeon game; a
submarine from a game
by Maxartists, and two in-
game mockup screens
from Clockwork Rian.

Sato
Takayoshi

Is pixel-based design art,
or more like a puzzle?
Takayoshi says developing
for today's cellphones is
similar to the early days
of 2D gaming.

Sato Takayoshi worked as a
character artist and animator on
the arcade to console ports of cult
2D shooting game *Sexy Parodius*
in 1996, as the only artist on the
team. He moved on from pixel
art after that project, creating
the entirety of the CG for popular
survival horror title *Silent Hill* on
his own. Here, he recalls what
it was like back in those days,
and it seems quite a bit like
the current state of cellphone
game development.

"When I was doing pixel work it
didn't feel like I was dealing with art
at all. It was more like a puzzle.
Basically, with the PSX and Saturn
I only had 15 + 1 colors (the last
one is transparency) for a pallet,
while the arcade hardware uses
32–64 colors on average.

Also, one "chara" file (256 x 256
pixels) wasn't big enough for the
character scale we wanted. Of

course the total number of chara
files we could use was limited (I
forget how many it was) so we had
to draw full animation sets using 32
or 64 colors and then divide them
into two or three colors and cut
them into tiny pieces in order to fit
everything in a 256 x 256 square
chara file [similar to the way people
do with mobile games now].

"One big character, like a level
boss, could be divided into 32
pieces of 16 x 16 pixels, 120 pieces
of 4 x 4 pixels, multiplied by two or
three pallets, multiplied by every
single animation frame. It was like
endless, really tedious, tiny puzzles
we were tackling. What's more, it
takes 15 hours of work a day, and
you feel like killing yourself.

"I think there will still be a place
for 2D art in the next generation
though. You need graphic design
for user interfaces, logos, insignia,
and things like that. You need

Sexy Parodius
screens

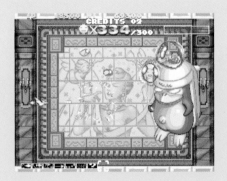

Mobile developers

Sexy Parodius was
a risqué sequel to
Parodius, which was
a parodic take on
Konami's popular
shooter, Gradius. It was
a straight-up shooting
game taken to a
deranged extreme.

concept art, production art, texture work—a lot of this is 2D. Two-dimensional graphic art will always be important. Even though we have 3D rendering media these days, most elements that make up a 3D image consist of 2D art.

"I don't, however, expect the old style of 2D game to become mainstream again. My wife's favorite games are still *PacMan* and *Mario Bros.* types of games, but these days that type of gaming experience is not

for everyone. I know I'd certainly prefer to play *PacMan* on the train than an FPS [first-person shooter].

"I admire the artists that still work in 2D though, like the *Castlevania* team. They produce stunningly beautiful art using only 16 colors. Producing great art within constraints has a lot of beauty to it. But we have far more complex rendering methods now, and there are much greater opportunities and challenges with 3D and other techniques. In this

industry have to keep taking advantage of these opportunities, or you lose your job.

"3D has greater dimension and depth, but it requires more and more people to create this sort of work. 2D has less expressive power, but you can control the entirety of the visual style and mood, and it can even make the game design a lot simpler. It is not rare for a cheaper, simpler product to surpass gigantic ones."

Sexy Parodius screens

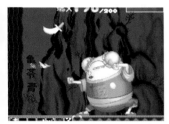 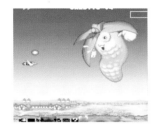  

Developer commentary: Sato Takayoshi

The game is, to put it bluntly, intensely strange. Converting the huge amount of images was no small undertaking.

113

Amateur hour: Marios 64

Michael McWhertor has created an alternate Mario universe, inspired by the genuine sprite's classic origins.

Michael McWhertor is fascinated by, and focuses on, video game art and marketing. He runs with a certain attitude, a consistent voice that doesn't always represent his own. Michael and his site were suddenly thrust into the spotlight when, after a fit of boredom, he put together "Marios 64": 64 Mario sprites, using the original palettes from the arcade *Donkey Kong* and *Donkey Kong Jr.*

The result is astonishing—64 Mario sprites that never existed in any game, but easily could have. With none larger than 14 x 21 pixels, and all sharing the same 16-color palette, these little Marios could each have come from their own game—bizarre as it might seem to imagine a game featuring Mario dressed like one of the Village People.

As Michael says, "I love pixel art and detest 99 percent of the 3D and line-art representations of many classic game characters. There is a certain purity to those flat, 16-color sprites that should never be forgotten." Michael started young: "*Super Mario Bros.* was played to death in our house.

I even thought *SMB2* was great!" This was a sentiment not always shared by players, and might speak of the depth of McWhertor's obsession. "I don't think I could ever tire of the character and level design from the *Super Mario Bros* series. I remember straining my eyes to extract every minute detail from Nintendo Power screenshots of *Super Mario World*."

There's a phenomenal range of subject matter represented in each tiny pile of dots. There are games (Ken and Ryu from *Street Fighter*, and Sega's *Alex Kidd* make appearances), and real people (all four Beatles) and plenty of pop culture (the Village People and a certain cartoon nuclear technician).

That these characters were created isn't as impressive as the fact they're recognizable, despite their limited palettes and diminutive size. And there are so many of them. It's a fantastic effort, made all the more surprising because no one's ever done it before. "I was just tired of thinking of Mario as a plumber and have always been fascinated by an alternate universe."

Mobile developers   If the Beatles were fat Italian plumbers with their own low-res, 16-color, 8-bit video game, they might look a lot like this.

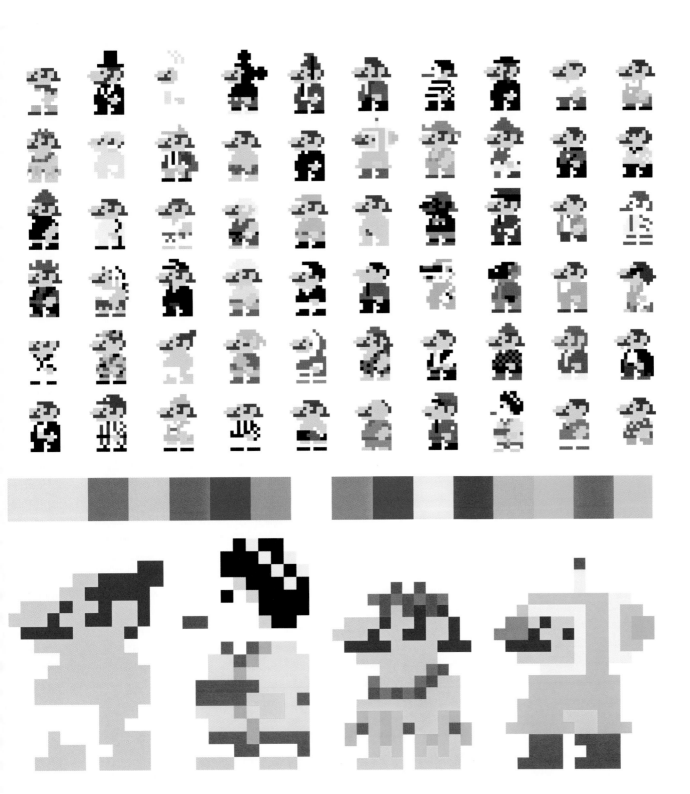

Amateur hour:
Marios 64

The variety by itself is
impressive. That such
variety could be realized
with so few resources,
and still be recognizable,
is incredible.

Creating graphics: PC Engine

Making games for old consoles is what Ian McPherson likes to do. Here he gives you his step-by-step guide to designing for the cult PC Engine system.

Creating games is a popular pastime for many people. A harder core bunch like to challenge themselves by making games for old consoles. The PC Engine has a loyal following around the world, and several coders are putting together new games for it. The last official game release was in 1999, but fan support keeps it alive.

PC-Gunjin, a play on the system's name, is being readied for a 2006 release. Graphics artist Ian McPherson is creating backgrounds and sprites for it. Like most other 16-bit systems of the time, the PC Engine had some very strict limitations on color palettes, sprites, backgrounds, and available memory. There were 512 colors to choose from, but only 16 colors could be used for each sprite or

8 x 8 background tile. Another limitation was that only 64 sprites could appear on the same screen. Ian had to be mindful of these things when making the graphics. Because of the limitations of the hardware, the boss opposite is made from a mix of both sprites and background tiles to achieve the desired effect. The following images detail the creation of a boss character, with commentary by Ian.

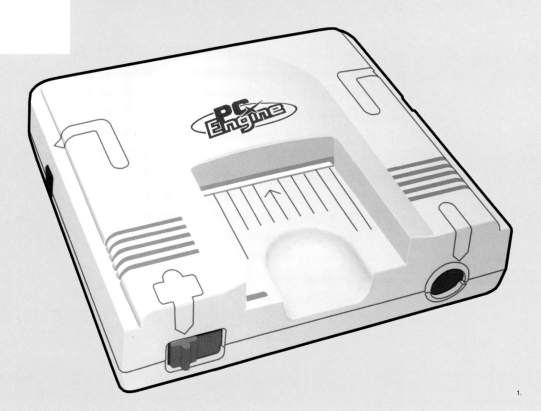

1.

Mobile developers

Creating games for dead platforms is not for the faint of heart. It takes a serious dedication to create games with such extreme restrictions. The end result, as these in-game screenshots prove, can eventually rival commercial productions released when the hardware was still fresh.

 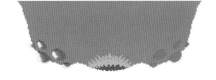 

1 Here, I started by making a green silhouette in the general shape I wanted the body to be. Most of my pixel graphics start out this way, as a base, and are built upon in parts or layers.

2 In this picture, I've added the eyes and the mouth. On the right you can see the purple silhouette I drew as a base, and on the left you can see the highlights that give them an almost jello-like look. The mouth was quickly drawn in as a reference, but was later changed.

3 Notice the full shading and highlighting applied to the eyes, and the new mouth. For the mouth I had wanted something very alien looking, and I think this worked out well. I also adjusted the overall shape of the body slightly to give the entire beast a more sagging-from-the-ceiling look.

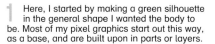

4 Here are the final tweaks made to the main body. I have clarified the light source on the green slime, and also contoured the outer edges. I know the light sources on the eyes are mirrored, but they did not stand out very much when drawn in shadow. The weak spot of this boss are the purplish eyes.

5 Now let's take a look at the mouth animation. I basically just used the original mouth as the base, and modified it from there. It's a simple three-frame animation—looks a bit like those eggs opening in the movie *Alien*.

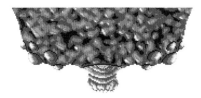

6 For the tentacles, I went with the tried and true 16-bit era, chain-of-ball-sprites technique. It's quite effective at giving the impression of tentacles, especially when you make each segment look interesting and tapering in size.

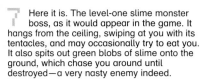

7 Here it is. The level-one slime monster boss, as it would appear in the game. It hangs from the ceiling, swiping at you with its tentacles, and may occasionally try to eat you. It also spits out green blobs of slime onto the ground, which chase you around until destroyed—a very nasty enemy indeed.

Creating graphics:
PC engine

Many of the techniques used to create this boss have been around as long as most people can remember. It's never easy to do, but advances in tools and techniques, built on the foundation of old games, makes it easier for amateurs to make their dream games.

Converting graphics

There's money to be made by releasing old games onto new systems. So what are the challenges of converting character designs from the platform they were designed for and onto a new one?

Jaded game players might consider it cynical when a developer brings an old game to a new console, and there's no doubt that some developers are perhaps too eager to shove the same game onto as many platforms as possible. From the developers' point of view, however, it would be crazy not to do it—development costs and risks are greatly reduced when a finished game is converted. The game is proven popular, the music and sound are finished, and the game dynamics are already polished. A small team is needed to make the old game work on new hardware, converting music and sound samples, reprogramming the game, and making the graphics fit the new screen.

Converting pixel art from one screen to another can be a huge problem. Pixel graphics are usually prepared with the specifications of the target platform firmly in mind. The artist is painfully aware of the limitations of the system, and how it affects his images. Everything about the system has an effect on the pixel art. Some systems had limited palettes, reducing the available colors. Some couldn't move large images, so individual objects had to be smaller. Systems with smaller overall resolutions would require a great amount of effort to make every onscreen object smaller, without using so much detail that they became unrecognizable. Older cartridge media would limit potential with their lesser storage capacity.

Creating sprites for the first time is an arduous task for any artist. The artist must convert sketches and drawings into pixel representations, and then do it again and again for each frame of a character's animation. The artist, after completing hundreds or even thousands of images, might finally finish placing these countless pixels and relax, only to find that

1. Sonic the Hedgehog
2. Sonic the Hedgehog

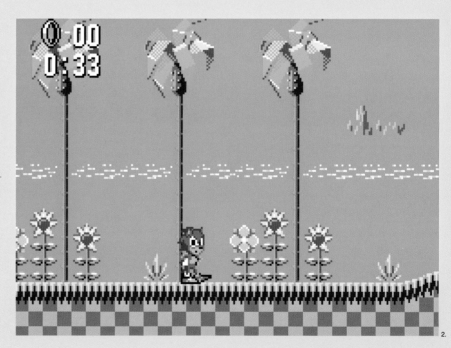

1.

2.

there's now a need to release the game on a different platform.

When a video game proves popular the publishers will invariably seek ways to cash in on the success. Licensing the characters to manufacturers of other media—such as toys and comics—is perhaps the most common way to do this, but many popular games will also see a release, called a port, on more than one game system. Porting a game will immediately make a wider audience available, but it also brings a new set of challenges to the developers and artists.

When porting to another platform, the existing sprites might need to be adjusted for size, color, and animation—a daunting task when the original sprite has been optimized for one device. If the new system requires a sprite to be shorter by two pixels, which ones should be removed? If they're clipped from the wrong place the character might suddenly find his nose blends with his face, or the tops of his shoes disappear. If he needs to be taller, where would the pixels best be inserted so the new design isn't detrimentally out of proportion?

In most cases the conversion is quite painless. In any given console generation no console will differ much from another. Similar color abilities and almost identical resolutions make the artist's job easier, but when the target platform is portable, with a significant loss of both resolution and color, the transition can be traumatic.

Consider Mario's leap from 1985's *Super Mario Bros* to 1989's *Super Mario Land*. When making the leap from the NES to the GameBoy, Mario lost four vertical and two horizontal pixels. The size constraints left Mario with a smaller eye, at least one less thumb, a much smaller mustache, and a hooked nose. Color changes were made too: Mario's hair became the same color as his hat.

1. *Super Mario Bros* (NES)
2. *Super Mario Land* (GameBoy)

1.

2.

Converting graphics

These two Mario images are from different generations of what is a very similar game. The NES and GameBoy *Super Mario* games were siblings, separated by massive differences in hardware ability.

Sometimes a developer will want to release a game on two different generations of hardware. It doesn't happen often, but while Sega's Genesis (MegaDrive) was topping the charts in North America and Europe, its older and less expensive Master System was dominant in countries like Brazil. Their big success, *Sonic the Hedgehog*, was so popular Sega ported the game backward from the 16-bit hardware it was designed for to the 8-bit platform that preceded it.

As you can see, Sega's artists managed to adapt Sonic for the older hardware without significantly impacting his appeal. The changes are immediately obvious: his shoes are the same color as his belly, there's much less shading on his head and spines, and the stripe on his shoe is almost completely lost.

This is in addition to his reduced stature—the older hardware required the ported sprite to be two-thirds as wide and three-quarters as tall as the original.

Scenarios like this are being played out over and over again in this modern age that finds most pixel games on portable systems with specifications very different than the consoles of old. In order to maximize

1. Sonic the Hedgehog, (Genesis 16-bit)
2. Sonic the Hedgehog, (Sega Master System 8-bit)

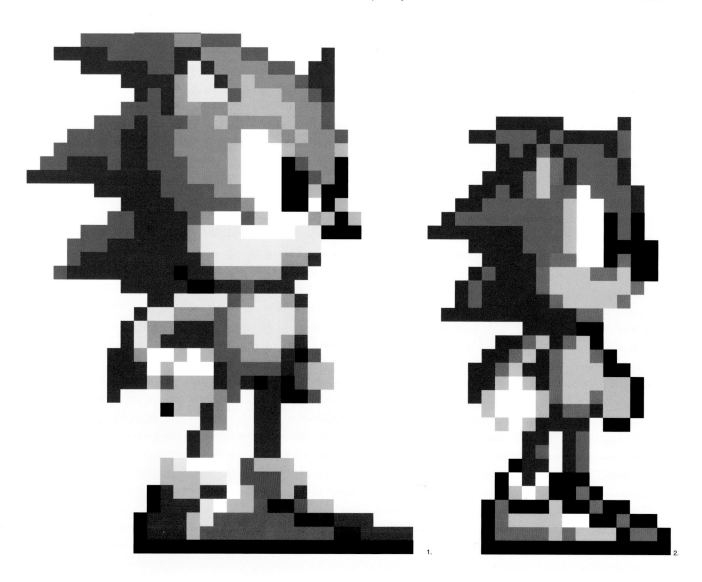

1.

2.

Mobile developers

Converting Sonic from the 16-bit Genesis to the 8-bit Master System was no small task. Everything was different—sprite size, screen size, color palettes, storage... Everything!

their resources many publishers are re-releasing old properties on new hardware. This is especially common on cellphones, where a recognizable game will stand out in a crowded market. To accomplish this, drastic steps often need to be taken, and it is the artists that bear the brunt of this work, since computers can't be trusted to do it on their own.

1. *Double Dragon* (Atari Lynx)
2. *Double Dragon* (Commodore Amiga)
3. *Double Dragon* (NES)
4. *Double Dragon*, Arcade

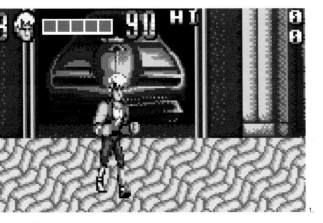

Converting graphics

These *Double Dragon* screenshots illustrate the extreme differences when porting across different platforms. The arcade source was fairly beaten into submission for the Amiga and NES versions, but the ancient Lynx managed to do it nearly perfectly, but for the total screen resolution.

Porting games to cellphones

Many games designed for other platforms are ported onto cellphones. What are the technical and artistic implications for character design and gameplay?

Most cellphone screens, except in Japan, offer very low resolutions and small screen sizes. Nokia's N-Gage, its flagship games platform, has a resolution of just 176 x 208 pixels. Worse, most mobile screens are oriented vertically, instead of horizontally. Consider Konami's *Castlevania* on the Super Nintendo, which had a resolution of 256 x 223. Compare this to the cellphone version, which had a mere 120 x 90 pixels into which the same game had to be squeezed.

In the image below left the sacrifices made are obvious. The cellphone version, while lovingly converted by I-Play, has seen the screensize reduced by almost two-thirds horizontally, and one-third vertically. The new screen layout is vertical too, which is often a challenging squeeze for a game designed to be played horizontally.

Creating new graphics for cellphones is a significant undertaking if a developer wants to reach the widest possible audience. Phones have a wide variety of resolutions and palettes, and the developer has to decide between creating graphics for the lowest common denominator or recreating the image for every different screen size. The alternative is worse, as some owners of hot new cellphones found their games ran in miniscule windows on their postage-stamp-sized screens.

Not all screens share the same aspect ratio either. Some phones have screens that have borders reserved for the phone functions while others use odd resolutions or aspect ratios. There is no standard, unlike for console games or portable game systems. When creating graphics for a

1. *Metal Slug* (cellphone)
2. *Metal Slug* (arcade)

1.

2.

Mobile developers

These screenshots of the mobile and arcade versions of *Metal Slug* manage to be very similar to each other, despite the awesome differences in platforms. The mobile version has limitations, but the spirit of the game shines through.

Nintendo GameBoy Advance or Sony PSP, the developer knows that every screen his game will be played on will have the same specifications as the rest.

It's a testament to the popularity of cellphone gaming that developers keep producing games against this kind of adversity. And pixel artists are so dedicated to their craft that they will keep resizing their pixels over and over again.

Modern phones are making this job easier for the developer. In the two examples below, Taito has managed to make a version of its classic *Bubble Bobble* that is so close to pixel-perfect it's unlikely that their artists were even involved. The programmers could simply reuse the arcade images in the new version, which speeded up the development process significantly. In the *Elevator Action* update, the graphics are actually better than they appeared

in the original. Advances in technology make the cellphone more powerful than the arcade machine released more than two decades earlier.

Converting games to mobiles is not likely a trend that will slow in the near future. What will be most exciting is platform standardization, so that every phone can play the same games. And a decent controller wouldn't be a bad idea.

1. *Bubble Bobble* (arcade)
2. *Bubble Bobble* (cellphone)
3. *Bubble Bobble* sprite
4. *Elevator Action* (cellphone)
5. *Elevator Action* (arcade)

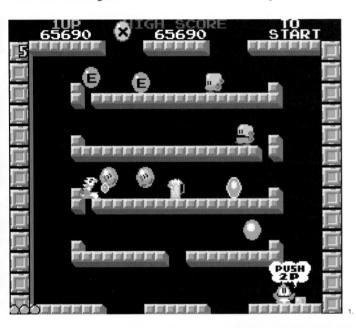

1.

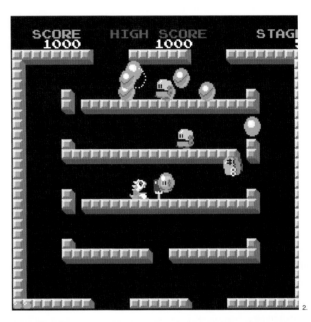

2.

3.

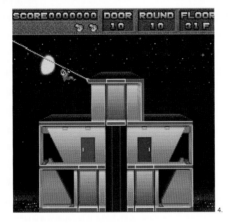

4.

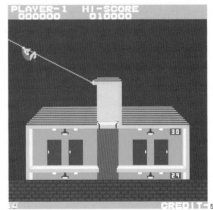

5.

Porting games to cellphones

The mobile version of *Elevator Action* is actually better than its arcade ancestor, with more colors, and a higher resolution screen. Mobile *Bubble Bobble* wasn't bettered, but it's very nearly identical to the original.

Porting SNES to GameBoy Advance

In the first in our analyses of porting games from one platform to another, we look at two powerful systems with some similarities, but many significant differences.

The GameBoy Advance saw more than its share of ports from the Super Nintendo. When the GBA was announced originally the common buzz was that the system was as powerful as a Super NES. You could almost hear the excited gasp from pixel pushers and developers around the world—finally, a portable system with some serious graphic capabilities. Pixel artists would now have a system that would allow them some creative freedom, and developers could start porting their Super NES properties to the new hardware.

The capabilities of the two systems weren't really that similar, but the approximate power was nearly the same. The GameBoy Advance was, for a time, the most powerful handheld game system available. Even though it lacked the dedicated game-playing chips that powered the Super NES, it had a fast multipurpose processor that could mimic these functions with relative ease. In fact the GBA was more powerful than the SNES, and could manipulate the pixels in ways the original SNES could never do. The cartridges also had a greater capacity than the SNES, allowing for richer and more extensive play.

This immediately opened the door not only to ports of classic SNES titles but also to enhanced versions of these old games. Nearly every company that had a hit game on the SNES rushed to port it to the GBA, and the pixels flew fast and furious. For a long time a disproportionately large part of the GBA library was comprised of SNES games.

Many of these games weren't direct ports. In most cases, the developers took pains to justify the second purchase of the game to players. Many included new game

1. Ghosts 'n' Ghouls (SNES)
2. Ghosts 'n' Ghouls (GBA)

1.

2.

Mobile developers

The GameBoy Advance, left, had far more storage for the game images than the SNES cartridge did, and all of the cut scenes got a fantastic upgrade in addition to a widescreen presentation.

1. *Yoshi's Island* (SNES)
2. *Yoshi's Island* (GBA)
3. *Shy Guy sprite, Yoshi's Island* (SNES)
4. *Mr. Nutz* (GBA)
5. *Mr. Nutz* (SNES)

modes, extra levels, and "arrange" modes, where the game levels were re-arranged to provide new variety.

Capcom's *Ghouls 'n' Ghosts* is a good example of this enhancing trend. The overall screen resolution is reduced, but the number of colors has been significantly increased. At least in the opening cinemas, the in-game graphics were left unchanged.

The average Super Nintendo game was 256 pixels wide, where the GameBoy Advance screen could display a maximum of 240 horizontal pixels. The big difference was in the vertical resolution—from 225 pixels to a very narrow 160. Most SNES games on the GBA became, by necessity, letterboxed versions. For some games this didn't cause problems, but with others the two-thirds vertical screen resolution meant cutting off significant portions of the screen. As you can see in *Yoshi's Island* and *Mr. Nutz*, some collectable items are nearly completely hidden with so much of the screen missing.

SNES to GameBoy Advance porting

Some games were ported with no differences at all. Both *Mr. Nutz* and *Yoshi's Island* were brought to the GBA without any significant changes. Screen resolution was decreased and visibility reduced, but the games are otherwise nearly identical.

Sega Master System to GameGear

Is the GameGear just a Master System in disguise? Porting games from one to the other is an exercise in playing to strengths and minimizing faults.

It's a little-known fact that Sega's popular color handheld, the GameGear, was a redesigned Master System. When Sega developers were looking for a GameBoy killer, they didn't have to look any farther than their own old console. It was a natural choice; the Master System was a mature technology, it could be manufactured cheaply, and would use a lot less power than newer, faster chips. The graphics were upgraded from a paltry 64 colors to a whopping 4,096. However, the exceptionally poor quality of the screen meant that these extra colors were wasted. Blurring was a serious problem, players were often killed by things they couldn't see, and the viewing angle was so strict that if the top of the screen was perfectly aligned the bottom would be faded and difficult to see.

Despite these faults, the Game Gear was very popular, and it had a surprisingly large library of hit games. This was largely due to the Master System's origins: the machine was fully compatible with older Master System games. Because of the larger palette, however, the reverse was not true: the GameGear could handle fewer colors, but the Master System choked on the new, larger palette data.

Both the GameGear and Master System were unusually fortunate. In Brazil, a toy company called TecToy was producing Master System consoles long after the platform had died off in other countries, while elsewhere Sega itself was promoting the GameGear. Unsurprisingly, huge efforts were made to keep the GameGear library fresh, and after Sega exhausted the good titles in the Master System back catalog, the company started looting the MegaDrive library. It's almost

1. *Sonic the Hedgehog* (GameGear)
2. (Sega Master System)

1.

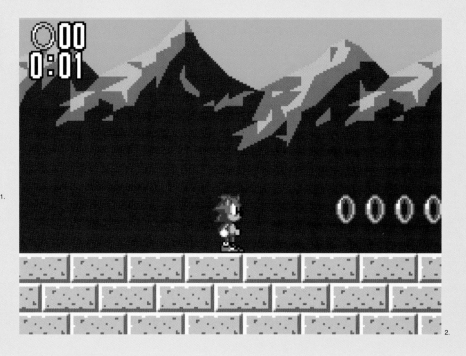

2.

Mobile developers

In most cases the GameGear games were simply cropped versions of the original Master System games, but when the GameGear version came first, the MasterSystem versions often suffered with empty spaces and slower action. The system was asked to draw more objects than it could handle.

unheard of to make games available for an older system after a new one has been released, but developing for a portable system changed all the rules. The Master System benefited in no small way from Sega's efforts to keep the GameGear viable.

Sega ported many games to the GameGear from the MegaDrive. Most games in the *Sonic the Hedgehog* series had a portable outing, and because of the large Master System market in Brazil, they were further adapted for the older console as well. Most conversions were easy, but they weren't all effective. The Master System screen resolution was much higher than the GameGear—many games ran slower on the Master System because the CPU had to move more

onscreen graphics than on the GameGear. This increased resolution also revealed more of the playfield, so Master System players had an easier time of things than their GameGear counterparts, who had less time to react when a bad thing entered the screen.

It's interesting to see how the conversion process differed. Games designed solely for the GameGear have a large black border on the Master System. Some games were designed on the Master System first, and have significant parts of the screen cut off. Others, designed for both, lose parts of the screen on the GameGear but generally did not lose anything important, as if the borders were merely decorative.

Sega's run with the GameGear was long but ultimately doomed, as Nintendo's GameBoy grew stronger and stronger. Eventually, the free ride came to an end, but not before the Master System had received far more support than it was due.

1. *Frogger* (GameGear)
2. *Frogger* (Sega Master System)
3. *Sonic the Hedgehog* (GameGear)
4. *Sonic the Hedgehog* (Sega Master System)
5. *Aleste 2* (GameGear)
6. *Aleste 2* (Sega Master System)

1.

3.

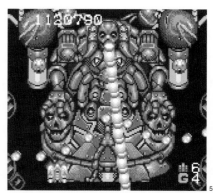

5.

2.

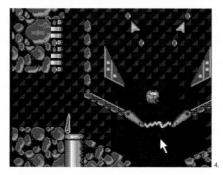

4.

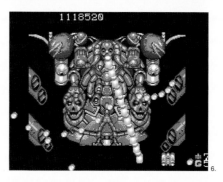

6.

Sega Master System to GameGear

Ghouls 'n' Ghosts

This legendary game was ported to many different systems. How was the game's development affected by the speed of technological change?

Capcom's *Ghosts 'n' Goblins* was an incredibly popular game when it hit the arcades in 1985. Despite a difficulty curve that bordered on pure evil it was a such a hit with players that it saw ports to most home computers of the era, and Nintendo's all-conquering NES. As a publisher Capcom was on the rise, as perhaps were most developers at the time. *Ghosts 'n' Goblins* was at the forefront of Capcom's surge of hit games to be released rapid-fire for several years.

The concept wasn't new—at heart it was a platform game like any other. The hero, Arthur, was an armor-wearing man out to rescue his girl from the clutches of an evil demon. The gameplay was tightly orchestrated—Arthur controlled well and the game was, even though unfair at times, easy to understand. The game had a joystick and two buttons: jump and

shoot, the norm at the time. The game's reputation was generated largely on the back of the graphics. The graphics, and especially the character designs, were beyond functional. Capcom excelled at creating graphics that were at once realistic and cartoonlike. Recognizable images littered every part of every stage, from vicious crows and fireball-spewing plants to winged demons, fearful because of their effect on Arthur's mission rather than their appearance.

Ghosts 'n' Goblins spawned a very popular sequel three years later. Imaginatively named *Ghouls 'n' Ghosts*, the new game offered a host of new features: new weapons, the ability to shoot up and down, golden armor, and very much improved graphics. Taking advantage of three years' improvements in computer hardware, the new game offered more colors, more and larger

1. *Super Ghouls 'n' Ghosts* (GBA)
2. *Super Ghouls 'n' Ghosts* (SNS)

1.

2.

Mobile developers

The Super Nintendo Super *Ghouls 'n' Ghosts* (left) was a visual marvel when it was released, with colors so lush and vibrant they rivaled arcade games at the time. When brought to the GBA years later, the game remained the same but cut scenes were improved.

1. *Ghouls 'n' Ghosts* (Arcade)
2. *Ghosts 'n' Goblins* (NES)
3. *Ghosts 'n' Goblins* (NES)
4. *Ghouls 'n' Ghosts* (Arcade)

enemies, special effects such as rain- and wind-blown trees, and a whopping 50 percent increase in resolution. There was more of everything, and the game looked sharper than most of its contemporaries did.

This new game wasn't any less difficult than its predecessor. It was so hard it would enrage players who, after spending countless quarters

in the arcade, reached the end only to find they had to start from the beginning and do it again—at twice the difficulty. But it looked fantastic at every step, and it played well. Capcom knew how to put a game together by this time, and the players kept coming back.

Ghouls 'n' Ghosts saw two console ports. A 6-megabit version was released for Sega's MegaDrive

(Genesis), and in Japan an incredible 8-megabit version was released on NEC's ill-fated SuperGrafx console. The SuperGrafx made very effective use of these extra two megabits, and even though the screenshots didn't show as many colors in the sprites, it had far more of the background effects that made the game such a joy to explore.

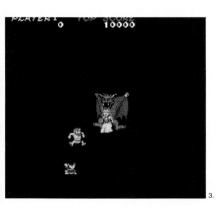

Ghouls 'n' Ghosts

The arcade version of *Ghouls 'n' Ghosts* (top left, bottom right) had a widescreen presentation and incredible animation. The NES port of the prequel was a little less awesome to look at, but no less fun to play.

Super Ghouls 'n' Ghosts

The final game in the series was *Super Ghouls 'n' Ghosts*, released on the Super NES in 1991. This additional three years saw, as expected, massive leaps in computing power. Nintendo's SNES wasn't the fastest machine in town, but it certainly had the muscle to astound players. The sound and graphics chips in the SNES made for an incredible experience, and Capcom pulled out all the stops to deliver what was one of the very best platform games ever made.

The resolution of this third game was the same as the first. Curiously, this wasn't noticed by most gamers until years later when emulation allowed comparisons between the two games. With CRT displays (old-fashioned glass TVs and monitors), a game system could stretch the display horizontally without a noticeable change in detail. When the SNES *Super Ghouls 'n' Ghosts* was released, the image, even though it was only two-thirds as wide, was stretched out to fit the screen. This allowed Capcom to make a smaller image—which was easier for the SNES to manipulate and required fewer resources—fit the entire screen. The end result was a game that

had the resolution of the first and the colors and sprite quality of the second.

Some of the sprites were recycled from the arcade version. Arthur himself was identical when jumping, but five pixels narrower when standing —a fact not likely to be noticed while playing. It's curious that Capcom didn't compress the jumping image as well.

The first and second *Ghosts 'n' Goblins* games were ported to nearly every popular console during their reign as two of the top platform games (the third was a SNES exclusive). Many home computers also received a port: the Atari ST, Commodore Amiga, and Sharp X68000, to name but a few. Some of these were amazingly well done. Capcom itself handled the X68000 port of *Ghouls 'n' Ghosts*, and it was a pixel-for-pixel perfect duplicate of the arcade version. Some of the others didn't fare so well—the NES port of *Ghosts 'n' Goblins* was a little anemic, and the less said about the Atari ST versions the better. An often overlooked semi-sequel was later released on the SNES in 1994, featuring the Red Arimer, the demon who caused Arthur so much trouble. Called *Demon's Crest*, it was similar

to Konami's *Castlevania*, as much a questing, semi-RPG as a platform game. The graphics in *Demon's Crest* were some of the very best to ever grace the Super NES, but the slower pace and lack of advertising combined with the new name killed its chances of long-term popularity.

The extra-cellular network i-mode created by NTT Docomo in Japan has seen a lot of Capcom properties brought over for cellphone play. *Ghosts 'n' Goblins* is one such game that received a full port.

There were several other interesting updates to the game, including a *Ghosts 'n' Goblins* RPG for the Japanese GameBoy, and a fun but simple minigame called *Ghost Trick* in SNK's NeoGeo Pocket Color game, *Millennium Fighters*.

1. *Ghosts 'n' Goblins* (Arcade)
2. The poor NES conversion looks surprisingly accurate in this screenshot.
3. The GameBoy Color version of the first game used sprites from the NES version, but the screen size was vastly reduced.

 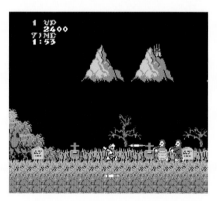 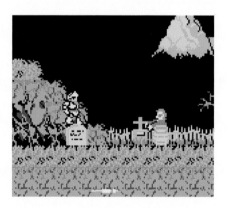

Mobile developers

These three versions of the game are remarkably similar. In the arcade version (left) there are more colors, but the NES version (center) offers a wider view on account of the smaller sprites. The GameBoy Color version was identical to the NES version, with a much smaller screen.

1. *Ghouls 'n' Ghosts* (arcade)

2. The MegaDrive version with its accurate palette.

3. The SuperGrafx version with the poor colors and excellent backgrounds.

4. *Super Ghouls 'n' Ghosts* was never released in the arcade, only on the SNES.

 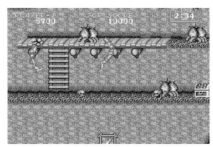 

5. There are no differences between this arcade screenshot (left) and the same game from the Sharp X68000 (right) computer version.

6. The SuperGrafx version is much more like the arcade version, but the MegaDrive had better color throughout.

7. When porting the GBA version from the SNES many changes were made—at least in the introduction. The game itself is identical, but for the smaller GBA screen.

Ghouls 'n' Ghosts

Ghouls 'n' Ghosts was a wildly successful game, reaching every major platform of the era. Each new platform had a version that played to its strengths and weaknesses, but as always playability was paramount, and no version disappointed in this regard.

Genre histories: RPGs

Character design and game genre go hand in hand, with each affecting the other and spawning design conventions and playing modes that still hold true today.

Perhaps more than any other genre, role-playing games (RPGs) have gone from some of the simplest graphics and designs to some of the most intricate.

Originating from pen-and-paper games, early RPGs were naturally sparse on the graphical designs and instead focused solely on replicating the experience of their unplugged counterparts. Graphics (if there were any besides text) were usually little more than stick-figure drawings, or were sometimes constructed from ASCII characters. Graphical elements at this point were basically just indicators; it could be argued that this was more due to the fact that technology in 1980 simply wasn't capable of generating the kind of images that would have been useful for an RPG, but there was also fairly obvious disinterest on the part of creators. RPGs were still very much a fringe genre and were often

coded by a single person or a very small team, so the graphical aspect became expendable in favor of the more traditional role-play elements.

This isn't to say there weren't sometimes brilliant attempts to infuse some artistry into the visual element of the games. *Akalbeth: World of Doom*, released in 1980, eschewed a text-driven adventure in favor of a crude (though impressive for the time) first-person dungeon crawl and an overhead map made using the then-powerful graphics capabilities of the Apple II computer. *Dungeons of Daggorath*, released in 1982 for the Tandy TRS-80 home computer, sported an improved, fully functional first-person interface and unique wireframe renditions for several monsters and items. Static monochrome, wireframe characters may not sound like a whole lot by today's standards, but when most games had little more than

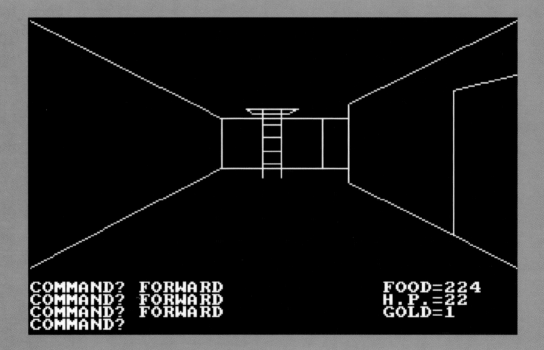

Genre / pixel histories:

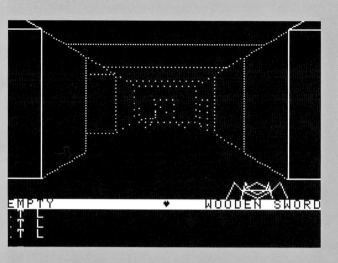

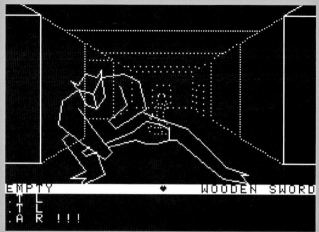

1. *Dungeons of Daggorath*
2. *Ultima III: Exodus*

placeholder graphics if anything, it was incredibly impressive, especially considering that *Daggorath* was a tiny 8,000 bytes in size. That's far smaller than most image files you can find on the Internet. Another influential early series was *Wizardry*, also for the Apple II, whose signature first-person head-on battle system would be widely copied by many console RPGs later. However, these were all exceptions to the general rule, and by and large the genre didn't move forward from a presentation standpoint for a number of years. The first computer RPG to truly

make notable forward strides in using graphics to help tell the story was *Ultima III: Exodus*, released in 1983. For the first time, characters were animated as opposed to static images, colors (all four of them) were all over the place, and the in-game interface was partitioned off using graphical elements. It was the first glimmer of what was to come in a few years' time; *Exodus* is frequently cited as one of the most influential computer role-playing games ever, and sure enough, its influence shows up almost everywhere in the genre you care to look. In Exodus one can see sort

of a basic prototype for all computer and console RPGs to come, but it wouldn't be until a few years later that the genre would receive its first true shake-up.

As computer role-playing games became more popular and began to garner larger cult followings, they were also gaining attention in an unexpected place: Japan. While RPGs had a social stigma as a closeted, geeky pastime avoided by the mainstream in the West, that sense of popular disdain was far less prevalent in Japan, where the concept was both fresher and more

exotic. Western-style fantasy games were quite novel in the mid-1980s, and that freshness led the company Enix to release the first installment of what would arguably become the largest console RPG franchise in history: *Dragon Quest* for the Japanese Famicom console in 1986.

Combining the tile-based overworld map from the *Ultima* series with the first-person Wizardry combat style, *Dragon Quest* took the first huge technological leap forward by being unafraid to reinvent itself where necessary. Gone were the stick-figure characters and heavy, stat-filled menus, replaced by a distinctly Japanese esthetic that streamlined the interface down to being navigable with the two-button Famicom controller. Where most

Western RPGs had long fallen into a deep rut of recycling the same Tolkein-inspired creatures and settings, *Dragon Quest* instead relied on its own world with its own population and rules base.

Ironically, this innovation may have had more to do with the subjective rarity of RPGs in Japan at the time. *Dungeons and Dragons* was commonplace in America and many computer-based games were clearly based on it, but in Japan, the Enix developers had to more or less wing it as they went along due to a dearth of Japanese language material that could influence them. This led to some interesting design decisions that would have never come out in the insular (and by that point rather incestuous) Western RPG

development scene. Notably, *Dragon Quest* was one of the first RPGs to actually give the player character a face. It wasn't much of a face, but was nonetheless a step in a different direction than most Western games would go. Previously, the heroes of the games were either never seen at all, or were mere stick figure representations, the idea being that the player was the character in the game so a fancy graphical representation was unnecessary. *Dragon Quest* did away with this notion by placing an emphasis on visual feedback, even to the level of a mild amount of alteration in the hero's appearance; equip the character with a weapon and shield, and the items would physically show up on his in-game avatar instead of merely being displayed as text on a

1. *Secret of Mana* (SNES)

2. *Secret of Mana* sprites (SNES)

1.

2.

Genre/pixel histories

menu screen. This subtle difference was indicative of the different design priorities in the West and Japan, and these would become increasingly more pronounced as time went on.

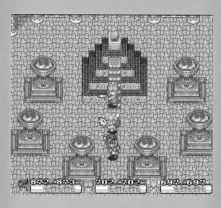

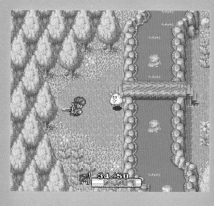

Genre histories:
RPGs

Dragon Quest was also one of the first games in the genre to sport an identifiable art style. Colors were vibrant and varied compared to the typical black-background American RPGs. Characters were drawn in the distinctly Japanese "super deformed" style where the head is usually the same size as (or larger than) the rest of the body, and enemy characters were often random and non-canonical in relation to other RPGs of the time. There were no elves, dwarves, or other fantasy staples, but all kinds of other creatures instead—golden golems, Shadow Knights, orange ghosts wearing witches' hats, and other creatures well outside the

norm for this type of fantasy world were commonplace in the Dragon Quest universe.

The ubiquitous Slimes—small, usually blue teardrop-shaped balls of goo with a face on them— became the iconic creatures associated with the series and have showed up in every single game in the line to date. While not the first console RPG to be released in the USA (that would be AD&D: Treasure of Tarmin for the Intellivison in 1982), Dragon Quest—renamed Dragon Warrior in the United States to avoid copyright issues—would become a smash success and would drag the RPG

genre out of its dusty niche market and into the ever-growing video game mainstream. Released for the NES, many gamers' first true exposure to the genre was through this game, and its influence in North America is arguably even greater than Ultima III.

However, it was also with the release of Dragon Warrior that the first rifts began to appear in the RPG genre—in terms of play style and design. Dragon Quest / Warrior is a reasonably "Western" RPG. While the creatures populating it are distinctly Japanese, the overall setting is close to other RPGs at the time. There is a vaguely

1.

Genre/pixel histories:

European medieval backdrop: magic works, scattered villages sell weapons of increasing quality (proportional to how far they are from the town the player starts in), and dungeons placed here and there have a big baddie sitting at the bottom guarding some nifty treasure. If you swapped out the Slimes and Golems for Orcs and Trolls, *Dragon Quest* could pass for a streamlined, overly bright Western RPG without too much trouble.

As technology marched forward, however, very pronounced differences began to arise between the American and the Japanese markets for RPGs.

While *Dragon Warrior* did well enough in North America, it was hugely popular in Japan, eventually launching RPGs into one of the most popular genres in that country. In direct response to that game's popularity, several Japanese companies began producing their own RPGs to cash in on the new craze, and the genre became much more Japanese in nature, eventually departing radically from its origins and becoming something totally different from what was being produced in the West.

1. *Dragon Warrior* sprite (NES)
2. *Dragon Warrior* (NES)

1.

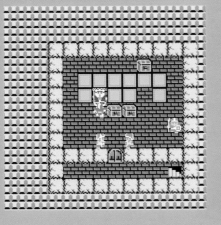

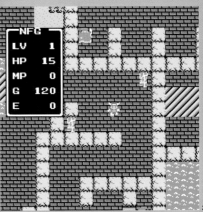

2.

There are several primary differences between Western and Eastern RPGs, but here we'll focus on esthetics.

Japanese games took appearance into account much more than their Western counterparts, and were typically much quicker to utilize newer technology to experiment with new visual themes rather than try to shoot for realism. American RPGs had almost uniformly spartan graphics when compared to other genres, so an action game for the Apple II would burn a lot more processing power on looks than a role-playing game. Japanese companies, however, quickly took to making their RPGs on par with other genres, often lavishing detail on enemy and location design to get across as cinematic an experience as possible. While an enemy in an action game might need to be

dummied down a bit to allow for things like animation, an RPG enemy was usually a static image that would show up in a battle window, so their sprites were often larger and more intricate than those that appeared in other games. This also allowed for distinctive art styles to develop even very early on. In this regard, one series in particular stands out.

Final Fantasy

A floundering Japanese company called Squaresoft would make history by taking a gamble with their own RPG, dubbed *Final Fantasy* as a company in-joke; if the game didn't sell, it really would be a final fantasy as the company would have to fold. Squaresoft had been working on several smaller RPGs for the ill-fated Famicom disk System, and *Final Fantasy* was to be the first (and presumably last) game to be

published on the core Famicom cartridge system. To say the game became a success would be a massive understatement. The *Final Fantasy* series has not only gone on to become the longest-running role-playing series in console history (including numerous spinoffs bearing the name) but is also arguably the most influential series ever after *Dragon Quest*.

A good portion of its popularity is due to its revolutionary esthetics. Square hired Japanese artist Yoshitaka Amano to create the hero and enemy designs, and then made faithful 8-bit representations of his signature flowing artwork and detailed character sketches. *Final Fantasy* sprites, even as early as the first game, have been among the most detailed and unique of any game in the genre.

1. *Dragon Warrior* (NES)
2. *Dragon Warrior* sprite (NES)

1.

2.

Genre/pixel histories: RPGs

The series was therefore quick to cultivate a specific look by doing simple things like arranging enemy sprites in action poses instead of having them simply standing there, using rounded-off tiles for maps as opposed to the right-angle filled tiles for the *Dragon Quest* games, and adding in even more customization to character appearance. Different weapons held by the player's characters would show up in-game, so if your warrior attacked with a sword, he would have an animation showing him swinging a sword. A hammer would produce a hammer, a whip a whip, and so forth. It would be some time before this was copied by other games.

Final Fantasy also took the step of making distinctive character designs, something that would become much more prominent as the series progressed. Taking a cue from Japanese anime, *Final Fantasy* characters would often sport

unrealistic hair colors, distinctive armor, or other unique identifiers to make them memorable and easy to pick out, compared to the often-anonymous or generic characters showing up in Western RPGs. As technology marched forward, these stylistic decisions became more and more complex and pronounced, and RPGs began to rise to the top of the heap as genuine graphical showcases, not only for the technology but also as a display of skill on the part of the developer. With the release of the then-powerful Super Nintendo, RPGs entered something of a Golden Age. While still not anywhere near as popular in the USA as they were in Japan, the stretch of time from 1992 to about mid-1995 saw the release of many of the definitive examples of pixel use in tile-based, 2D role-playing games.

1.

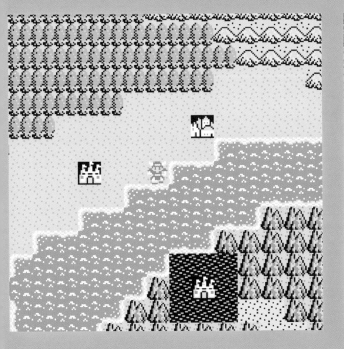

2.

Two games in particular are excellent examples of how far the medium was being pushed: Square's *Final Fantasy VI* (released under the name *Final Fantasy III* in the USA) and *Chrono Trigger* stand as a pair of the most remarkable and forward-thinking games released at the time. They have held up amazingly well, and these games still look excellent even by today's standards, thanks to a strong artistic and esthetic consistency. They pulled the rare trick of both pushing the boundaries of what was technologically possible at the time while still maintaining solid art direction to ensure that they would age gracefully.

This is one of the pitfalls when it comes to creating graphics, pixel-based or otherwise. Anything that appears new now will very quickly look old, and if the only reason something looks good is because it's at the cutting edge, it's not going to stay there for very long. Anyone

heading into the development end of these games nowadays would be much better off making something consistent to the vision of the game in general. These three games are some of the best examples of that. They also illustrate the benefit of having characters created beforehand by genuine artists. Even though all the primary actors in a pixel-based game are eventually going to be pared down to little more than a few colors on a grid, it's still possible to keep the individuality and expression intact, as these games were able to prove.

Final Fantasy VI was the last game in the series in which the characters would be designed by Yoshitaka Amano. Fittingly, the game would also sport graphics that adhered most closely to his original concepts. Every character was dramatically different in appearance than the other, not only in terms of gender or hair color but in every way.

Color pallets for the principal cast were wildly different while being easily identifiable at a glance, stealing a page out of old superhero comics. Boiling main characters down to a few primary colors will cement them in the minds of viewers more quickly. For example, most people see blue and red and instantly think of Superman, green and purple evoke Hulk, and dark blue and yellow represent the original Batman designs. Likewise, the primary protagonist of *VI* had distinctive green hair and red clothes, while the other primary female in the party was a bottle-blonde wearing all white. All of the characters, even in their small on-screen personas, were crafted as a distillation of the highly detailed character sketches they were designed from.

Another large innovation on the part of *VI*—and one that Square had been using for a few games but

1. *Final Fantasy VI* (SNES)
2. *Final Fantasy VI* sprites (SNES)

1.

2.

Genre/pixel histories: RPGs

which reached its pinnacle here—was making a uniform set of actions and expressions for each character: standing, walking, crouching in pain, surprised, lying down, preparing to attack, waving a finger, glancing to either side, head bowed, laughing, and raising both arms.

Since all the characters were far too small to have any facial animations, these full-body actions served the same purpose and were used liberally throughout the game, both in battle, in general conversation, and in any other scripted physical action. It seems like a simple thing, but Square was able to get some amazing mileage out of just that limited action set by using it in creative ways (and using it everywhere it was possible to). Take, for example, "glancing to the side." On a functional level, this seems pointless. All characters are drawn facing four possible directions already, and it would be easier to simply have them face the relevant direction rather than draw a separate sprite of them with their body facing forward toward the screen while turning only their head to the side. However, just that little bit of effort paid off immensely in terms of what could be done emotionally with the

in-game sprites. Characters could now shake their heads to indicate "no" by rapidly alternating between glance-left and glance-right; they could walk toward the bottom of the screen in pairs while facing each other to have a conversation (a very natural act that would otherwise be impossible to represent visually); they could whisper asides to each other; appear to avert their eyes if embarrassed, or seem to peer at someone from a distance.

For example, imagine a game character walking toward the bottom of the screen. Let's say he's walking through a town and sees someone at the end of the street he recognizes, or maybe a pair of people he knows who are having a conversation. Traditionally, the game would be programmed so the first character would be walking downward in center screen, and then might stop to say something like "hey, that's such-and-such over there, and he's talking to what's-her-face. I wonder what they're talking about?" The camera then pans downward to the pair in question, who are facing each other, and lets the player hear what they're saying.

1. *Final Fantasy VI* sprite (SNES)
2. *Final Fantasy VI* (SNES)

1.

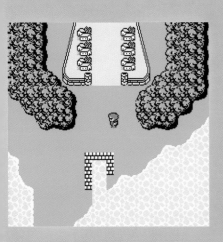

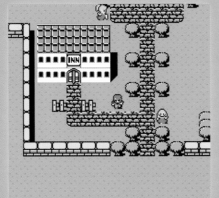

2.

Now imagine the same scene, but using the head-turned sprite. The first character is walking down the road in the town, and then sees the same pair, but to his left or right, and turns (while still facing forward) to look at them. Then, the pair on whom the player gets to eavesdrop wouldn't have to be facing each other directly, but could be walking along having a much more discreet-looking conversation.

All this subtlety results from one very simple alteration to the regular forward-facing sprite. It may seem like an incredibly minor detail, but it's exactly these sorts of details that will add a whole new level of polish to a game. While many players won't explicitly notice small details like that, they are still processed on a subconscious level and add an

important level of life to a games presentation. It's important to note that this applies only to characters in the game that are actually going to be displaying emotion, however!

By stark contrast, compare the character design in *Final Fantasy VI* to that of the enemies encountered during the game. Unlike the player's characters, enemy sprites are intricately detailed static images that are much larger and properly proportioned, a design tactic used since the very first game in the series. The reasoning is pretty simple. Let's say the party is going to fight a huge dragon. The player is already perfectly familiar with the characters in his own party, and provided they're given the proper range of emotions, is likely already quite attached to them. The dragon

the player is fighting, on the other hand, is not a character he's going to attempt to empathize with, but rather an enemy he's going to try to defeat in combat. The dragon isn't going to need to look sideways at anything convincingly. It just needs to sit there and look cool while it and the player wail on each other, so at this point it's a much better idea to sacrifice mobility and multifunctionality in the dragon's graphics for sheer brute force. Of course, as a side-effect, the dragon needs exponentially more pixels to make him look more detailed, which means he needs to be physically larger the more detailed he gets. This is beneficial in the long run, of course, as you're pretty much killing two birds with one stone: fighting a big, detailed dragon is a lot cooler than fighting a small boring one. At first, it might sound as

1. *Chrono Trigger* sprites (SNES)

1.

if the two styles would jar together—a realistically drawn dragon fighting a bunch of misproportioned, comparatively plain party members. However, this is almost never the case.

It's an interesting detail, and perhaps one of the most important, yet frequently overlooked, aspects of any games visual design. Graphical realism is very different from graphical believability, and a good art team knows to shoot for the latter over the former. Even though tile- and pixel-based games obviously have nowhere near the realism of, say, a next-generation 3D rendered game with all the bells and whistles, it's not at all a stretch to say that many people will identify with the pixel graphics much more easily.

In short, as long as a games visual design is consistent, it need not be realistic. In fact, that can be strongly to its detriment. None of the characters in *Final Fantasy VI* or *Chrono Trigger* look even remotely realistic, but because of the smaller details lavished on their design, they come across as more believable than they would have as 3D characters. A player's mind is perfectly willing to fill in the blanks as long as the characters are able to emote convincingly in some form or another. One of those sprites making an exaggerated, surprised look by throwing their arms over his head and sweating will go make a much stronger connection with the player, which encourages more involved, more personal gameplay.

In a way, it's very similar to how a cartoon works. Many cartoons, especially those hailing from Japan, have main characters that are markedly less detailed than their surroundings. This is because a

simple design on the characters makes them easier on a psychological level for a viewer to project themselves into. As a general rule, the more realistic a character looks, the more distanced he is from the viewer and the less empathy he will inspire, a trick that goes as far back as the earliest Disney cartoons—the good guy always had a big, relatively blank expression. He had large eyes, exaggerated movements and facial features, and a lack of overall detail, while a villain would have small eyes and sharper, less inviting features, and was likely to be much more realistically proportioned.

When you carry this time-honored principle to video games, then the stylistic choice made in the *Final Fantasy* series makes even more sense; it's the exact same idea, and it just so happens that pixel graphics lend themselves well to it. It would

be a serious problem to animate several small expressions and positions onto a super-intricately detailed player character sprite, but it's not nearly as much of a problem to do so to the comparatively disposable enemy sprite. This has the added bonus of working well on a subconscious level for the player.

Now, if *Final Fantasy VI* is one of the best examples of character-based sprite graphics, then *Chrono Trigger* shows off how to do environments within a tile environment and make them every bit as intricate and stylistic as 3D graphics. Unlike characters, pixel-based environments are at their best when they show an insane attention to detail. The primary key is not to slip into repetition, a huge pitfall when you're dealing with creating a world that's made up of repeating tiles, and *Chrono Trigger* is full of examples of how to do this. Right away, the level of detail in the world becomes apparent, with items such as furniture tiles looking as though they were built into the location instead of simply inserted from a pool of tiles. While *Final Fantasy* would have objects such as a chair tile, a bed tile, or a desk tile, Chrono took the next step and incorporated full-on additions to an environment that looked like they were supposed to

be there, carrying on the inspired move toward subtlety seen in other Squaresoft games and bumping it up a notch.

In an interesting precursor to full-blown pre-rendered backgrounds, Chrono would also pull simple cinematic tricks like repositioning the camera from the standard RPG three-quarters overhead view to the classic sidescrolling 2D view for some scenes. Other shots were prerendered completely, letting the design team move outside of the confines of the tileset to create unique locations for game events.

Of course, all of this can backfire horribly in the hands of a lazy or uninspired development team, but by the same token none of these tricks are particularly difficult to pull off. Creating a few minor frames of motion for a character or creating a single unique background or two is only marginally more difficult than

skipping it, but the overall effect on a game's esthetics can have huge repercussions. Small details are what amount to a graphical spit-polish for 2D RPGs and are usually just enough to suspend a player's disbelief completely.

So to bring this concept together, an overarching theme is to have easily distinguishable, heavily animated main characters capable of displaying emotion while still leaving enough "wiggle room" in their overall looks for player superimposition and empathy. Make your environments detailed and unique, and spend all your remaining graphical horsepower on intricately designed enemies for the player to fight.

For a number of reasons, RPGs never really matured to the same level graphically as their Eastern counterparts, though a very convincing argument could be made that the actual gameplay of Western

1. *Chrono Trigger* sprite, Square (SNES)
2. *Chrono Trigger*, Square (SNES)

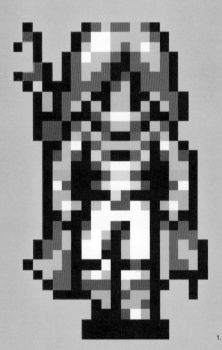
1.

2.

RPGs evolved on a far more significant level than most Japanese variants achieved.

This isn't due to a lack of artistry on the part of Western designers so much as a difference of focus. While in the West the emphasis was on player choice and immersion, Japanese RPGs were increasingly designed around telling a specific story that was locked in from the very beginning, almost always starring specific predefined characters as opposed to player-generated ones. Thus, a Western RPG would need to be more flexible than an Eastern one by sheer virtue of the fact that the player was given far more choice about whom to control. It was often more practical (not to mention technologically feasible) to leave many of the details of a character's appearance up to the player's, rather than the designer's, imagination.

From a strictly graphics-oriented standpoint, this makes Japanese RPGs far more relevant as a case study, a fact that many modern-day Western RPG companies are taking into account. Since the technology — not to mention the consumer base—now exists to blend open-ended Western RPG gameplay

styles with the polished image more common in Eastern games, it's becoming increasingly common for the line between the two to be blurred. Thanks to the proliferation of the Internet, there is also a greater readiness on the part of both East and West-based consumers to accept either style. On top of that, most current-day creators of RPGs were raised on a mix of both styles to some extent, and this blends the styles together even further. The trick, of course, is to take the good aspects of both. No one says an RPG can't look like it comes from Japan but play like it's from the USA. It's just that very few people people have tried to make such a game, particularly during the period when cellphone platforms were not yet powerful enough to pull them off well, but consoles were too powerful.

Given the popularity of cellphone platforms nowadays, however, it's likely only a matter of time before someone steps up to the plate and uses examples from both regions to create a truly unique hybrid RPG that blends the best of both worlds. Given how well the genre formats itself to a handheld market and the near-limitless creative possibilities it offers, such a game could be impressive indeed.

1. *Chrono Trigger* sprite, Square (SNES)
2. *Chrono Trigger*, Square (SNES)

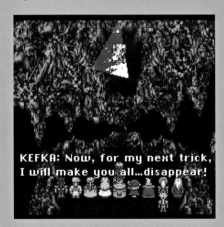

Genre histories: Fighting games

Fighting games don't pull their punches when it comes to iconic character design...

The fighting game genre didn't emerge out of nowhere, but it certainly erupted onto the scene. For a long time, head-to-head fighting games fuelled a massive gaming boom, and companies like Capcom and SNK built empires around them. Both companies—along with many other less successful ones and yet more hangers on—drove hardware development for a decade.

One of the very first fighting games was Data East's 1984 dual-joystick experiment, *Karate Champ*. It was a complicated affair with very simple graphics but a surprising variety of moves. Two opponents would meet in front of a mean-looking judge and, using both sticks, maneuver back and forth, trying to land the winning blow. One hit was enough for the judge to raise his little flag and announce a winner. It was simple and not much to look at, but very popular.

The player characters were recognizably human, were well animated for the time, and looked as good as they could with only 13 colors.

A year later Konami created what was almost the first modern fighting game. Konami introduced many new features that later became staples of the genre: status bars that decayed as a player took damage; a roster of opponents, including two hidden ones that had to be defeated in sequence; and the gravity-defying leap. There weren't many moves, but the character art was superbly charming. Unlike *Karate Champ*, however, it didn't offer head-to-head play, and so the fighting game boom was delayed a little longer.

Capcom's *Street Fighter* was a foreshadowed revolution. Released in 1987, it was almost

1. *Street Fighter,* Capcom (Arcade)

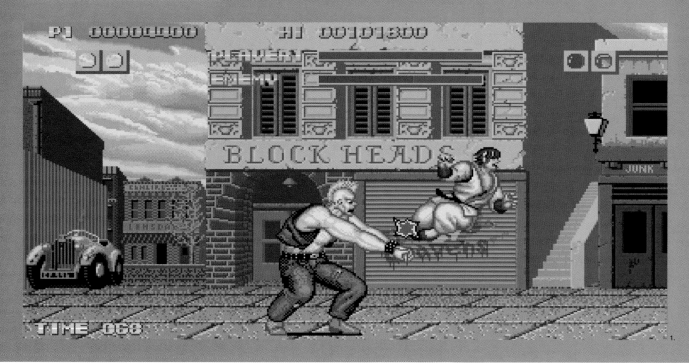

1.

Genre/pixel histories:

The original *Street Fighter* wasn't a patch on its successors. It wasn't really popular, it wasn't that good looking, and it didn't play well. Its place in history was assured simply by being the game that spawned *Street Fighter II*.

exactly right, with almost 100 colors on every screen and some characters that had shed the primitive look mandated by older hardware. The animation was choppy because of the limited graphics storage available, and the characters were unattractive, but the graphics were more realistic than in previous games. While popular, it didn't cause quite the sensation that its sequel did.

Nearly four years later Capcom released one of the most influential games ever made. *Street Fighter II* was a sequel light years ahead of its predecessor and of Capcom's competitors. Featuring over 200 colors, new animated backgrounds covering a wide range of locales, and eight playable characters, *Street Fighter II* was an unqualified triumph. The colors lent it a rich vibrancy that previous games lacked, and each character had a wide range of movement, including special moves.

Capcom had got into its stride. Other companies entered the fray, and the fighting game genre exploded. Every new Capcom game was better looking than the last, with more memory devoted to newer and better graphics. Capcom's competitors, most notably SNK, didn't hold back either, with rapid-fire releases of every variation imaginable. By the time *Street Fighter II* was released, onscreen characters had reached over 100 pixels in height, more than double 1984's *Karate Champ*, and by 1998 they had grown so large they threatened to reach the edges of the screen.

1. *Vampire Savior*
   Capcom (Arcade)
2. *Street Fighter*,
   Capcom (Arcade)

Genre histories:
Fighting games

Capcom was, for a very long time, the undeniable king of the fighting game. Every new game was bigger, faster, and more excessive than the last.

2D fighters remain one of the strangest, yet also most consistent, genres in all of video gaming. Here is a genre with almost no dependence on modern hardware, that still based primarily in arcades, using what many would assume to be a hopelessly outdated design model to appeal to an ever-more-enclaved audience of hardcore fans.

So, perhaps more than anything, it's a testament to the brilliance of the original concept that 2D fighters are not only still around but they're still remarkably popular. Many fans of the genre will play them to the exclusion of nearly everything else, and the subculture has spawned countless Web sites, underground magazines, and even a movie. Few gamers take their games as seriously as fighter fans, which can make it an incredibly difficult genre to break into, but ultimately the fighter audience can prove to be the most fiercely loyal bunch around for those companies that still "get it."

Here, we'll take a look at what shaped the genre into what it is today, and how—even within such an apparently limited framework—many fighting games have truly pushed the boundaries, marrying simple gameplay with a unique esthetic that has won the genre its legions of diehard fanatics.

Most game historians will agree that the Arcade coin-op *Yie-Ar Kung Fu*, by Konami, is technically the first example of a 2D fighter. Released in 1985, the single-player game starred a nameless martial artist who would fight his way through two tournaments, facing his opponents (most of whom were named after the weapons they used) one at a time. With only two buttons and an eight-way joystick, control was limited, but the general framework for a fighting game was already in place. The player moved about in real time and tried to hit his opponent while avoiding his opponent's attacks.

1. Fatal Fury, SNK (NeoGeo)

1.

Genre/pixel histories

SNK's *Garou: Mark of the Wolves* was a visually incredible evolution of one of its flagship games, *Fatal Fury*. The style was significantly different, and the animation was liquid smooth and glorious to behold.

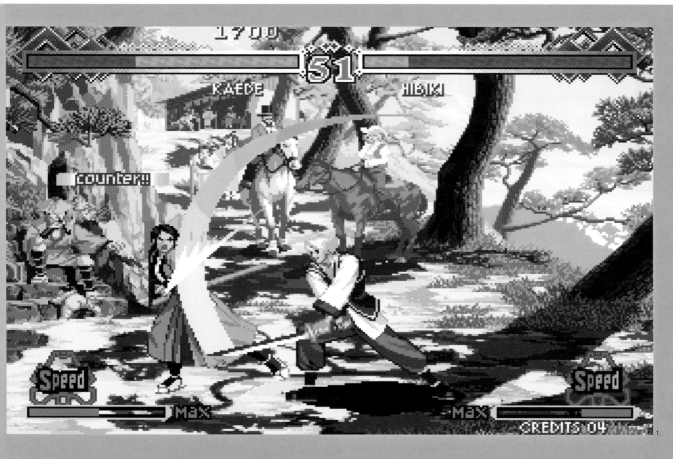

1. *The Last Blade 2*, SNK (NeoGeo)

Each fighter had a "life bar." A successful hit against the player's opponent would decrease his bar, while getting hit did the same to the player. After defeating the computer-controlled opponent the player advanced to the next level to fight a more difficult opponent.

The game wasn't a particularly huge smash hit but was popular enough to spawn a spin-off—Capcom's *Street Fighter* of 1987. *Street Fighter* was a two-player game where the player had to fight off a number of ever-more-difficult foes. While the game could be played by two people, they didn't get any choice as to which character they could be. Player one took control of Ryu, a Japanese fighter with brown hair in a white "gi," while player two took control of Ken, who was simply a color-swapped version of Ryu with a red gi and blond hair. Both players had identical moves.

A few twists this time around helped elevate *Street Fighter* into a genuinely popular arcade game, however. While still technically a two-button game (one for punch and one for kick), the buttons used for *Street Fighter* were actually pressure-sensitive, so hitting the punch button harder would result in a slower but more damaging attack. This was possible up to three levels for punch and kick, effectively giving the player six different base attacks.

There were also a number of special moves requiring a particular joystick motion followed by an attack button, though these moves were notoriously difficult to pull off. On the other hand, a single special attack could eliminate a third or more of an opponent's life bar, so they were worth learning if the player hoped to get far on a single quarter.

Genre histories:
Fighting games

Also by SNK, was a short-lived fighting game that never disappointed in any category, especially the visuals.

Needless to say, anything that encourages players to hit an arcade machine as hard as possible is just begging for trouble, and *Street Fighter* machines began to break down with such frequency that the game's setup was eventually replaced with a six-button layout that allowed players to choose which attack they wanted to use without having to bang the machine quite so hard. This six-button layout remains the hallmark of Capcom fighters.

The graphics in the game were also fairly advanced for the time, showing off Capcom's powerful arcade hardware, the CPS-1, or Capcom Play System One. While Ryu and Ken were fairly generic looking, both of their environments and their opponents were often strikingly detailed. The background stage would change from fight to fight, offering panoramic views of a medieval castle, the Great Wall of China, and other landmarks.

While the original *Street Fighter* may have been the original and was fairly popular for its time, it wouldn't be until the sequel, *Street Fighter II: The World Warrior*, was released that the genre would reach the form it's known by today. This all boils down to a single design element that subsequently became the entire focus of the game, would create a massive craze in arcades nationwide, and altered the course of fighting games forever. Not only could players fight each other simultaneously, they could now choose to play as any of the game's eight characters.

While Ken and Ryu returned for the sequel (and retained an identical move list), the remaining six characters were wildly different from each other, both visually and from a

1.

Genre/pixel histories

gameplay perspective. In retrospect, it's the characters from *Street Fighter II* that have become the most recurrent fighter styles, showing up in some form or another in virtually every single 2D fighter released since. You have a large grappler with short range (Zangief); a fast female character (Chun-Li); a tough American military character (Guile); the "weird," usually non-human fighter (Blanka); the skinny fighter with long reach (Dhalsim); and the big-but-speedy midrange fighter (E. Honda). Each character had a distinctive personality on display, with even Ken and Ryu starting to drift apart in terms of personality and onscreen presence. Ken's signature two-finger victory sign and huge grin upon winning a fight were markedly different from Ryu's stoic gaze into the distance.

To say *Street Fighter II* was a success would be a massive understatement. Almost overnight, a massive watershed community built up around the game's competitive play. Capcom responded by churning out several updates to the original game, adding in new characters, tweaking the gameplay balance to make the fighters more equal with each other while allowing for more strategy, and generally polishing up the mechanics of the game as much as possible. During most of these tuneups, however, the characters themselves remained the same, appearance-wise.

While it wasn't to the detriment of the game's underlying mechanics, it didn't take long for the characters' lack of personality to become apparent, especially in light of a competitor that was trying to muscle in on Capcom's business by taking the fighter stereotypes Capcom had established and turning them on their head.

1. *Street Fighter II*, Capcom (Arcade)

While Capcom was watching the money roll in from its *Street Fighter II* machines, other companies were watching as well and plotting to take a slice of this new, extremely lucrative head-to-head competition market that had sprung up almost overnight. The problem was, Capcom had hit a home-run in terms of play mechanics; there was very little about the *Street Fighter II* combat engine that could be done better, so the focus shifted from trying to outdo Capcom at their own game to trying to do something Capcom hadn't done yet.

When arcade veteran SNK set out to make its own fighting game, *Fatal Fury*, in late 1991, the producers took the extra step of creating characters that were not only well outside the established stereotypes, but had genuine personalities of their own. Unlike *Street Fighter*, the *Fatal Fury* fighters weren't blank slates for the player to project themselves onto, but fully formed characters with histories and distinctive personality traits, all of which were animated into their onscreen personas.

It was still a little rudimentary at that point, but while Capcom was busy retooling the engine behind a single game, SNK was slowly but surely building an entire universe for its characters to inhabit, and their games soon became synonymous with stylish graphics and memorable, unconventional characters who made their Capcom counterparts look downright staid by comparison.

This is due in no small part to SNK's new arcade hardware, dubbed the NeoGeo MVS, or Multiple Video System. Most NeoGeo cabinets could hold four games in them at the same time. A player would insert a quarter and could then select which game of the four they wished to play, which not only gave the player more choice but acted as a cost and space-saving measure for arcade owners, as they no longer had to buy a separate dedicated cabinet for each of the increasingly prolific SNK releases. This meant the cabinets were very popular with retailers,

1. *Samurai Shodown*, SNK (NeoGeo)

1.

Genre/pixel histories

Samurai Shodown was a game that catapulted SNK to success, and clearly shows the progression of SNK's artists. By the time the series finally ended the NeoGeo hardware was doing things no one could have imagined when it was launched.

giving the games a significant amount of exposure. In addition, the hardware itself was extremely advanced for its time and could handle a truly staggering amount of content on the screen, far more than Capcom's CPS systems. The compromise was that all NeoGeo games ran at a fairly low resolution of 304 x 224 pixels, but the tradeoff was worth it when you saw the games in motion. Great care was taken, particularly on flagship games, to add in minuscule details to characters wherever practical in order to breathe as much life into them as possible.

Here are some prime examples of SNK's craftsmanship, broken down by character.

Iori Yagami

Making his first appearance in *King of Fighters '95*, Iori would come to symbolize the kind of design that made SNK famous. Sporting a modified Japanese school uniform top, a bright red hairdo swept over one eye, and wearing red leather pants bound together at the knees (which oddly didn't seem to hinder his fighting ability), Iori was unique from the get-go. What truly sold the character were the little touches added to him, which have only proliferated over the course of the many games he's appeared in. Unlike most other fighters, Iori rarely ever throws a closed-fist punch, preferring instead to scratch, headbutt, or pinch and tear chunks of flesh from his opponents. He has an almost snakelike quality, with a tall, thin body and a perpetually hunched stance. A powerful kick from Iori actually has him using his entire body; he gets down on one hand and pushes his body, feet first, into his opponent's knees. While projectile attacks are common in 2D fighters, Iori again stood out by throwing a gout of purple flame along the ground, rather than the more common airborne projectile. Countless small embellishments round him out when the fight is over, anything from leaning over and taunting his fallen opponent, to pointing at the moon in the sky, to ripping one of the cuffs off of his sleeves with his teeth in frustration, should the match end in a time-out.

Hinako Shijo

One of SNK's many "playing against type" characters, Hinako is a small, thin, blonde girl in a purple dress with a matching hat, who fights like a Sumo wrestler. A good example of how creative animation can make an otherwise ordinary-looking character become extremely memorable, Hinako will toss salt on the ring at the beginning of a match, raise and stamp her foot Sumo-style, and has a whole run of Sumo-based moves, from a hand slap to a heaving toss.

Yuri Sakazaki

Few characters in any fighting game have as much personality as Yuri. A spunky Japanese girl wearing a relatively plain outfit of a white shirt and dark sweatpants, Yuri ranks as one of the most elastic characters in the entire SNK roster. She frequently parrots other fighter's moves and has a particularly pliable facial expression at all times, rare in a 2D fighter, most of whose actions are body movements.

1. *Samurai Shodown*. SNK (NeoGeo)

Screenshots like this one caused many gamers' hearts to race when they first appeared in magazines.

Genre histories: Platform Games

From Donkey Kong onward, platform games gave rise to some classic characters that rapidly became design icons.

The platform genre was, for a time, what made or broke a home console. Platform games—which can loosely be described as any game that features navigation of platforms of various heights as a core gameplay element—were the perfect showcase for mascots, which were used to brand consoles (and even entire companies, in the cases of Nintendo and Sega). The platformer has since lost much of its relevance in that regard, but continues to be a compelling genre, although it no longer pushes the 2D graphical boundaries it once did.

The genre began humbly. Nintendo's *Donkey Kong*, released to arcades in 1981, was one of the first notable platformers, and was the birthplace of Jump Man, who later became Mario. He had to navigate a series of slanted platforms, climb ladders, and jump over barrels thrown by Donkey Kong in order to save the princess

at the top. The arcade board couldn't handle many colors and the game was rather slow, but it was still advanced for the era.

This was the beginning of the one-screen platformer, in which games had to be cleared one screen at a time. Another notable game in this arena was Data East's *Burgertime*, a 1982 arcade release that challenged a chef to construct giant hamburgers by walking across the ingredients, causing them to fall. He had to crush enemies (condiments) along the way. The objects onscreen were larger than those in *Donkey Kong*, and far more detail was added, specifically to the burger buns. Still, this was a rather slow game. It wasn't until 1983 that platformers were able to consistently match the speed of earlier games such as *PacMan* and *Space War*. *Lode Runner*, released to computers such as the Commodore 64 (and later, to

1. *Super Mario Bros* (NES)
2. *Yoshi's Island* (GBA)

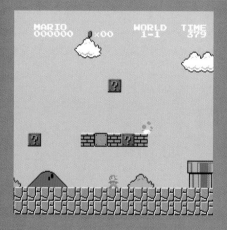

dedicated arcade boards) was incrementally faster than previous platformers, though it sacrificed color depth and detail. Also in 1983, Nintendo released *Mario Bros.*, which perfected the physics of platform-jumping in a one-screen environment, with a previously-unseen level of graphical detail in terms of characters. It also featured some early pixel shading to indicate a third dimension in the platforms.

The next great innovation in pixel platforming graphics came in 1985, with *Super Mario Bros.* for the NES. Up until that point, the majority of graphical innovations came from the arcades, but platform games were intrinsically suited to consoles, so from then on, most important evolutions in platform graphics took place on home consoles.

Super Mario Bros. was the first major platformer to feature scrolling levels—2D maps that the player could navigate horizontally and vertically. Graphics had finally evolved to the point where

characters could look moderately like their illustrations, but the 8-bit graphics were still quite limited. Sega's Master System also deserves a note, and was host to Sega's original mascot, Alex Kidd. However, it wasn't a significant improvement over the NES until much later in the lifetime of the hardware, which was continually revised, perfected, and picked apart by Sega's TecToy division in Brazil. After the advances made by the NES and SMS, the next major issue to tackle was color.

The PC Engine, or Turbo Grafx outside of Japan, was an NEC-created home console which only had an 8-bit processor, but boasted a 16-bit graphics chip capable of displaying 512 colors—and larger sprites than the NES. As the first 16-bit home machine, it had the biggest and boldest graphics for a while, and its seminal platformer was 1990's *Bonk's Adventure*, which featured blazing speed, large sprites, and deep, vibrant colors.

Sonic the Hedgehog though, released for Sega's Mega Drive/Genesis in 1991, was more widely regarded for its graphical advances. The speed of the game was unparalleled, and it featured hardware-supported parallax scrolling, which allowed multiple layers of background to scroll at different speeds, adding the illusion of 3D depth. Sonic became the defining character for Sega, and a mascot for all of the company's systems to come. Further iterations of the series were more about detail and expansion of level design than anything else in the graphics department, and the series was also notable for designing some of the first maps that were as tall as they were wide, making smooth, multi-directional scrolling a must. *Sonic 2* featured incredible depth of field via the parallax in the intro scene, and featured some of the brightest, most vibrant colors yet seen in games.

1. *Super Mario World* (SNES)

Then along came Nintendo to shake everything up again. With the release of the Super Nintendo, with its superior graphics and sound, the graphical face of games was to change once more. *Super Mario World* introduced new and wondrous graphics techniques to the console world, in the form of transparency, scaling and rotation, due to the console's "mode 7" processing power. True sprite transparency was quite new in home systems, having previously been approximated by dithering effects on the Sega and NEC consoles. Rotation and scaling were also most often fudged, so the effortless effects found in this game were awe-inspiring at the time. The huge characters, bright colors, and effects all made for a convincing package. This was the new face of 2D platforming.

Not content to stop there, Nintendo, through then first-party developer Rare, pushed the boundaries even further with the release of *Donkey Kong Country* in 1994. This was one of the first platform games to feature prerendered graphics, giving the game a pseudo-3D look, even though the game played entirely in the second dimension, and the resultant images were still sprites. The fact that the images were created and animated to appear 3D, then forced into 2D sprites made for compelling visuals for some, and Sega used this technique. This technique still saw sparing use in 3D games for a number of years (*Crash Bandicoot* is an example), for objects that didn't actually need to be rendered as polygons.

It was Sega who then took this technique to its logical conclusion, making pseudo-3D environments in 2D. *Sonic 3D Blast* was an isometric platformer, a bit like *Marble Madness* with jumping, released in 1996 for the Genesis and Saturn consoles, and was entirely 2D, though it operated in a 3D-like environment. That is to say, there were three axes of movement, due to the tilted perspective, but everything in the game was composed of prerendered sprites. The intense focus on animated backgrounds and vibrant characters kept this experiment from total failure, but it was not that well received by fans, who were eagerly awaiting the first true 3D rendition of *Sonic the Hedgehog*, given that the 3D-capable Saturn console was already on the market.

This was also not the first attempt at isometric design for the *Sonic* team, with *Sonic the Hedgehog Arcade* predating *3D Blast* by some three years. *Sonic Arcade*, though, was hand-drawn through and through, so it was *Blast* that put all of the new techniques together in one package. A much more lush entry into the field of isometric gameplay was Climax's *Landstalker*, also for the Genesis.

1. *Donkey Kong Country* (SNES)

1.

This game, though technically an action RPG, saw the majority of its gameplay center around platforming, and had very advanced topography for a non-3D game.

Another stopgap before the true 3D era was Sega's 32X, an add-on to the Genesis/Mega Drive that allowed for rudimentary 3D, but which was largely used for slightly better-than-Genesis level 2D graphics. *Knuckles Chaotix* was the best platformer for the console, and featured software driven transparency, and true scaling and rotation for the first time. Unfortunately, this only put the Genesis' 2D capabilities just about at the SNES level, and Sega turned its collective eyes toward the future. Also in the obscure 2D camp came *Rayman*. While the game itself isn't unknown, the best version largely is, as it came out for the ill-fated Atari Jaguar. The game's 2D graphics were quite detailed and incredibly well animated, and the Jaguar's dual

32-bit processors handled the strain like no other contemporary console could.

By now, the era of the polygon was steadily marching on, and sprite-based games were taking a back seat quite quickly. Even so, some games continued to roll out, proving that the 32-bit era meant better 2D as well as markedly better 3D. Astal was a particularly nice platformer, and came very early in the life of the Sega Saturn. Released in 1995, it boasted huge sprites, impressive scaling, and deep colored backgrounds. The Saturn though, like the Genesis before it, was incapable of transparent effects by default, which limited the presentation somewhat.

The PlayStation on the other hand, did not have this problem. Transparency was a cinch for the console, but its visual memory was much smaller. So it was a bit of a

trade off, which was most easily seen in cases like *Castlevania: Symphony of the Night*. Whereas the Saturn version boasted more animation, more enemies and a larger map, the PlayStation version had transparency, and crisper 3D elements. Other games followed suit, with *Megaman 8* and *Silhouette Mirage* facing similar issues on both consoles. Both games had their own appeal though, and as sprites grew ever larger, and animation more detailed, some hope was held out that the 2D platformer would continue into the future.

Genre histories:
Platform games

Not to be outdone, Nintendo released the cartridge-based Nintendo 64, which was a much more 3D capable system than both the PlayStation and the Saturn, but was also host to a few 2D platformers as well. The first was *Mischief Makers*, developed by Treasure. The game featured an odd combination of prerendered and hand-drawn sprites, and an entirely 2D flat perspective. The sheer number of onscreen characters possible was impressive, and the use of scaling and transparency was quite good, but ultimately the game wound up looking messy, and did not advance the pixel medium significantly.

The second 2D platformer, though— *Yoshi's Story*—was a prerendered tour-de-force. For those who like prerendered graphics—and it is an acquired taste—*Yoshi's Story* was at the top of the list. At the same time, Midway chose to release an action/platforming version of its *Mortal Kombat* series for both the PlayStation and N64, which featured 2D digitized characters—one of the last times any company would ever

do so. While the game was rightfully quite poorly received, the digitized technology had been perfected to the point that the characters looked quite unique and interesting, without being as blocky as they had been in the early *Mortal Kombat* and *Pit Fighter* days.

Operating entirely peripherally to the entire console market was SNK's arcade/home game line which featured some of the most detailed 2D animation, some of the biggest sprites, and also some of the worst slowdown at times. The *Metal Slug* series was particularly notable in the platforming department, not for its innovation in gameplay, as countless titles had come before it in the "run and gun" platform genre, from *Contra* to *Gunstar*. Because of the arcade board's ability to stack on as much RAM as the company desired, the animation was superb, and the games had a retro quality to them no matter how contemporary they actually were.

Unfortunately by now, 3D had captivated the hearts and minds of mainstream gamers, and 2D pixel art

was relegated to handhelds, which up until recently couldn't do proper 3D without great difficulty. Many of the handhelds had console equivalents, graphically. The Game Boy Color was similar to the NES. The Game Gear was similar to the Sega Master System. The Turbo Express in effect was the Turbo Grafx. The Game Boy Advance was similar to the Super Nintendo. As such, the graphical advances on these handheld consoles, at least in terms of 2D, were incredibly similar to the advances on each individual console that preceded it. The Game Gear was a bit of an anomaly here, in that the graphics of the console seemed more of a bridge between the Master System and the Genesis, providing Genesis-like speed and sprite size, with Master System-level color and scaling capacities. That *Sonic the Hedgehog* and *Gunstar Heroes* held up well on the handheld proved its worth well enough.

The Game Boy Advance was probably the most refined of this era of handhelds though, with the largest library and the most natural hardware-assisted sprite pushing

1.

power. Recent Treasure hits *Gunstar Super Heroes* and *Asto Boy* pushed the boundaries of how much the system could handle in terms of sprite numbers, explosions and speed, all without causing undue slowdown. The GBA is near the end of its lifecycle as well though, and Nintendo's DS has taken up the 2D reigns. The system is more robust than the GBA, with roughly the same power as the N64, but the 2D art has not advanced very significantly, given the declining popularity of the graphical medium in general. *Castlevania: Dawn of Sorrow* is a particularly nice platforming example of the DS's pixel-pushing prowess, and even reuses some sprites from the 32-bit *Symphony of the Night* title. While it appears as though

the DS will continue to host 2D platforming games, the genre is not advancing graphically in the way that the fighting genre is, for example, with *Guilty Gear* and *Fist of the North Star*, both high-resolution, well-animated hand-drawn games from Arc Systems Works.

It seems as though for the time being, the platforming genre won't be advancing pixel art from a technological standpoint any time soon, at least not in any official capacity. In the amateur scene though, interesting things are afoot. The amateur development scene is strong in Japan, and while the level of 2D may not be quite up to standard in the platforming genre, the fighting games created by

amateurs are often impressive. At the same time, looking forward isn't always the best way to advance an art form. *Cave Story*, created by an amateur artist/game designer who calls himself Pixel, has a purposefully retro feel, combining the old-school esthetic of the NES and Genesis eras, with the power of modern PCs, to create a very charming and unique world, full of vibrant characters and puzzles. Here, the pixel is not seen as primitive, but rather as a way to express a simple artistic elegance rarely seen in today's complex, jumbled gaming landscape.

1. *Metal Slug*
2. *Castlevania* (DS)
3. *WarioLand (Super Mario Land 3)* (GBA)

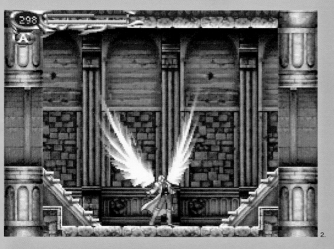

Mario sprite history

From Jump Man to hero and even movie star, Mario defined for many what was achievable in character design with just a few points of light onscreen.

There was a time when Mario was the most recognized character in the world, ahead of even Disney's Mickey Mouse. When Nintendo's *Super Mario Bros. 3* was released in 1990 there weren't many kids who would fail to recognize the rotund plumber. He wasn't always so recognizable however, and in fact his earliest origins gave little clue to his eventual appearance.

It's rare that a character spans so many hardware generations. Very few games have the longevity that Mario has had. Most are based around a gimmick and burn out within two generations, but Mario endures. Mario has, at last count, graced more than 10 Nintendo platforms. From the very earliest appearances in the arcade where he was known only as "Jump Man" to Nintendo's latest platforms, Mario has been a Nintendo staple. Throughout this long history, Mario has gone through many changes.

As the hardware progressed and as Nintendo's artists developed their skills, Mario grew and changed, often in dramatic leaps.

In 1982, Mario—or Jump Man— was astonishingly good-looking considering the limitations of the hardware. It's a famous and oft-told story that Mario's appearance was initially dictated by hardware limitations. Low resolutions and primitive video hardware made very strict demands on what Nintendo's artists could achieve. The first Mario sprite was made from only three colors—red, blue, and flesh. Mario wore coveralls so the movement of his arms would be apparent: a blue arm on a blue shirt would be unseen. He was given a blue mustache (to match his blue eyes and hair) to separate his nose from the rest of his face, and he wore a hat because hair was hard to draw.

1. Mario sprites 1981
2. Mario sprites 1983
3. Mario sprites 1985
4. Mario sprites 1989

1.
2.
3.
4.

Genre/pixel histories

Mario rushed through his evolution. From humble beginnings to superstardom there were rarely times when a newer, better Mario wasn't entertaining players somewhere in the world.

Sprite 1

Mario's appearance was also dictated by his function. In his first appearance in *Donkey Kong*, Mario only jumped over things, so he has the appearance of a cartoon hurdle jumper. When he appeared in *Mario Bros.* he was given—along with five extra colors—the new ability to smash into the floor above him, so his new look was more vertical, with a raised fist. This was to remain the same for almost all subsequent *Mario* titles.

Sprite 2

In 1985's *Super Mario Bros.*, the first of the modern Mario games, Mario's appearance was downgraded a little from the arcade *Mario Bros.* The new hardware, the Nintendo Entertainment System, didn't have the ability to replicate the arcade sprites. It was three colors again for Mario, although with the help of some rather unusual mushrooms he could now grow to double his original size.

Sprite 3

Super Mario Bros. 2 was the next big Mario outing, at least outside Japan. This new Mario had a different appearance as a result of improving skills among Nintendo's artists. Mario is no longer seen from a side-on perspective, and is now more rounded and cartoon-like.

Sprite 4

The GameBoy *Mario Land* was released next, in 1989. Unlike the recently released *Super Mario Bros. 2*, the Marios of *Super Mario Land* have an almost retro style. This gray Mario most resembles a squashed version of the original, and the larger, mushroom-powered Mario is nowhere near twice as tall.

Sprite 5

The unusually colored Mario from the second *Super Mario Bros.* game was discarded in 1990 when *Super Mario Bros.* 3 came out. The blue outline is now black, and Mario's orange skin is exchanged for a more appropriate color. Gone are the whites of Mario's eyes, and back again is Mario's one-handed sky-punching pose.

Sprite 6

When the Super Nintendo was launched only a year later Nintendo needed a "killer app" to fight off Sega's conquering Genesis, and Mario provided that app. The new hardware allowed far greater freedom for the artists creating the sprite, but the expanded color palette seemed to cause some trouble at first. Mario's colors are blotchy and don't blend well.

1.　2.　3.　4.

5.　6.

Mario sprite history

Nintendo made some odd choices along the way. *Super Mario Bros 2* was not originally a Mario game at all, and why is he outlined in blue? The SNES *Super Mario World* saw a Mario marred by poor color choices as Nintendo's artists struggled to come to grips with the new hardware.

In 1992 another GameBoy *Mario* was released: *Super Mario Land 2.* This new game featured sprites that were most like *Super Mario Bros. 3* in appearance, continuing a tradition of the GameBoy graphics from the previous generation. Nintendo had given up on replicating the entire *Mario Bros.* experience with the GameBoy this time around, instead zooming in on the action so that the parts of the screen that were visible were more recognizable. The new large Mario was almost twice as tall as the first GameBoy Mario.

This was a time when Nintendo was leveraging Mario like never before, using the power of the brand to make the NES to SNES transition as well as making the GameBoy more attractive against Sega's recently

launched GameGear. New Mario games were coming out every year, and 1992 was no exception— Nintendo re-released four (and later five) Mario games on one SNES cartridge: *Mario All-Stars.* This next change for Mario was at the same time one of the largest and smallest changes. Nintendo had come to grips with the expanded palette the SNES offered, so these games, identical to the originals, used the same sprites with far more colors. The resulting Marios are, for many, the definitive ones, possibly as good as pixels could possibly get.

In 1994 Nintendo released a fantastic GameBoy offering, *Donkey Kong,* but despite having the same name as the 1981 arcade game, this one featured Mario in an action

puzzle environment instead of the straight-up platform action of the original. It was a great tribute to the original and included all of the bonus items from the original game (the hat, purse, and umbrella dropped by Mario's ladyfriend) and, as the first four of over one hundred stages, all of the original arcade levels.

The march of hardware and Nintendo's skills allowed for some very real progress. The below left image shows Mario, from 1991's *Super Mario World,* on the back of *Yoshi,* his dinosaur friend. The image (top right, below) shows a baby Mario on Yoshi's back again, from the 1995 game *Yoshi's Island.* Even though the action in both games is strictly two dimensional, the new sprite isn't at all as flat as

1.

Genre/pixel histories

As time went by, and Nintendo's artists grew accustomed to the little plumber, Mario changed for the better. Happier, friendlier, and better looking in all his incarnations.

the first, showing far more detail and character with only four extra pixels' width.

This is where Nintendo lost the plot. At the same time they were making the transition to 3D with their Nintendo 64 platform, Nintendo started experimenting with different styles. Their attempts at pre-rendered Mario sprites, made from 3D models flattened for 2D use, resulted in the unattractive Mario you see below left from the 1996 *Super Mario* RPG. This prerendered technique was brought back for the 2004 release of *Mario vs Donkey Kong,* a fantastic game marred by a protagonist that's a nearly

unrecognizable pile of pixels. Nintendo proved it wasn't always easy to make attractive sprites, even when the subject matter was as proven as Mario. *Wrecking Crew '98'* and *Mario and Luigi* both featured Mario sprites that weren't a patch on their predecessors, despite more experience and newer and more capable hardware.

Super Mario Bros. 2
Originally this game was not made by Nintendo and didn't feature Mario at all, but pressed for a Christmas release, Nintendo slapped some new sprites into it and sold it as *Super Mario Bros. 2*. It had a different gameplay dynamic:

pulling plants from the ground and throwing them at enemies replaced jumping and smashing blocks. Despite this shift Mario still raised his fist, in fact both his fists, when he jumped. Interestingly, Nintendo made one small change to the sprites when they re-released the game in 1995: Mario no longer raised his arms when he jumped. Unless Mario was carrying something the same sprite used when he was running was used when he jumped, making it look like he was gliding.

1.

Mario sprite history

At the end Nintendo, for reasons known only to the company, pushed Mario a pixel too far. The last 2D Mario sprites are much less attractive than in the SNES era.

Donkey Kong sprite history

One of the first iconic game characters, Donkey Kong helped establish game character design as a multibillion-dollar industry.

Donkey Kong, originally a love-struck ape trying to find a little peace with a kidnapped maiden, didn't go through nearly as many style upgrades as his nemesis Mario. First appearing in the arcades in 1981, *Donkey Kong* was originally a game designed to reuse the hardware from the poorly selling *Radar Scope*.

Shigeru Miyamoto designed both Mario and Donkey Kong, and while his struggle to create a good-looking plumber with such limited resources is well known, he didn't seem to have as much trouble with the monkey. Donkey Kong is much larger than Mario, allowing for greater design freedom. Even made of only three colors, Donkey Kong is immediately recognizable.

There were only minor changes made between his debut in 1981 and his return appearance in 1982, as the caged father in *Donkey*

Kong Jr. He was made of different colors this time, but was very much the same monkey.

In 1983 Donkey Kong saw his first real revision. *Donkey Kong 3* had a very different monkey on the title screen than the one seen in-game. He was rounder, more cartoon-like, with much smaller legs, larger eyes, and plus-marks in his teeth that are familiar to anyone who played the Nintendo games of the era. In-game Donkey Kong looked very similar to the first two games, though he'd lost the jowls and had a very different facial expression.

That was the last we saw of Donkey Kong for a very long time. It was eleven long years before Donkey Kong appeared in another game, but the comeback was astonishing. Nintendo made a very extreme makeover with two all-new games in 1994.

1. *Donkey Kong Jr* (Arcade)

1.

Genre/pixel histories

Mario's nemesis came a long way, though for a while it looked like he would be caged, shelved and forgotten as Mario hogged the limelight.

The GameBoy was home to a new Donkey Kong game, confusingly with the same name as the original arcade version. The new game was a hybrid sequel to the original, blending some aspects of the classic game with gameplay taken from *Super Mario Bros.* Once again, Nintendo fell back on the old damsel in distress theme, and Mario, armed with a new arsenal of moves, had to navigate a hundred fiendish puzzles to rescue the girl. Donkey Kong, despite being the feature character, was relegated to a a barrel-throwing boss every four stages. Despite looking as good as he could in four shades of gray, the 1994 *Donkey Kong* was known more for the near-perfect gameplay than the graphics.

1. *Donkey Kong Country* (SNES)
2. *Mario vs Donkey Kong* (GBA)
3. *Game & Watch Gallery* (GBA)
4. *Donkey Kong* (GameBoy)
5. *Donkey Kong Jr* (Arcade)

Donkey Kong sprite history

Donkey Kong never received the love that Mario did. Until the SNES *Donkey Kong Country* games were unleashed, it seemed that the giant ape was doomed to languish in obscurity.

Donkey Kong Country for the SNES was a revelation. Developed by the Nintendo subsidiary, Rare, it was one of the most effective uses of pre-rendered graphics ever. While Sega was bringing out new hardware to bolster the aging MegaDrive (Genesis), Nintendo used the superior palette of the SNES to deliver brilliant, realistic graphics.

Never mind that the graphics were originally drawn on expsneisve Silicon Graphics workstations, *Donkey Kong Country* wowed everyone. Each element in the game, from the smallest animated sprite to the immense backgrounds, was painstakingly created out of polygons. Using the most

sophisticated tools of the time, the graphics were pre-rendered, so that the expensive computers did the hard work of converting polygons to flat images that the SNES could manipulate like any other sprite.

This was a real turning point, both for Nintendo and the entire game industry. It paved the way for some truly dreadful graphics as machines with limited palettes were forced to animate computer-generated sprites designed for computers that had, for all intents and purposes, no palette limits at all.

Gone were the days of painstakingly edited pixels and handcrafted sprites. Massive workstations

could crank out sprites that, at first glance, looked and animated better than any hand-made sprite. Gone also was the ape Miyamoto made, replaced instead by a hip, attitude-wielding CG ape.

Donkey Kong Jr. suffered a less disastrous fate than *Donkey Kong* did. From the 1982 *Donkey Kong Jr.* game it was another 16 years before Junior saw another game Nintendo's *Game & Watch Gallery 2* for the GameBoy Color, released in 1998. He was a very different monkey by this time—rounder, cuter, and the very model of a modern ape.

1. *Donkey Kong* sprites (SNES)

Genre/pixel histories

The pre-rendered look pioneered by Rare for the SNES *Donkey Kong Country* games looked great in motion, but didn't age particularly well. Unlike handcrafted sprites, the computer-generated pre-rendered look harsh and over-contrasted.

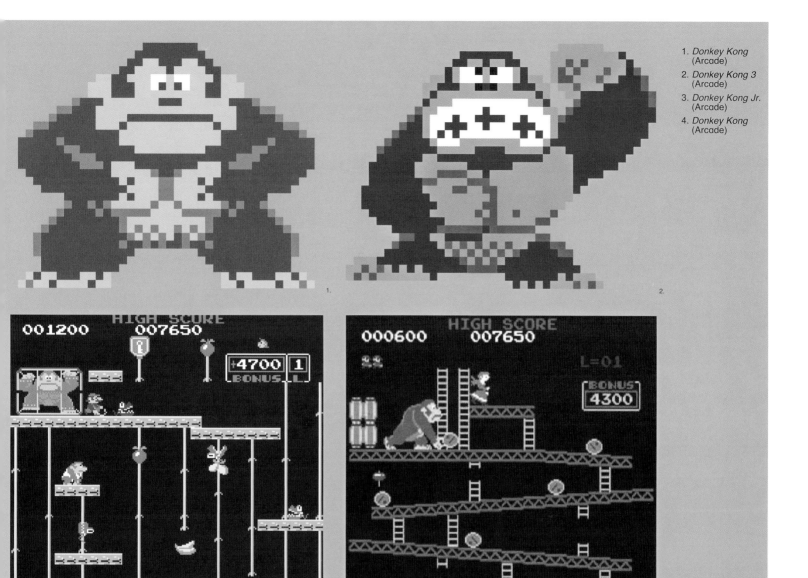

1. *Donkey Kong*
   (Arcade)
2. *Donkey Kong 3*
   (Arcade)
3. *Donkey Kong Jr.*
   (Arcade)
4. *Donkey Kong*
   (Arcade)

Donkey Kong
sprite history

Donkey Kong didn't
create the platform
game genre, but it
might as well have.
In the arcades in
1982 it was the
undisputed king.

Sonic the Hedgehog sprite history

In the opposing camp, Sonic became Sega's mascot, and a global brand to rival a certain animated mouse in the movies. Again, pixel artists transformed a few points of light into a world-famous face.

Sega's *Sonic the Hedgehog* didn't go through as many evolutionary stages as Nintendo's Mario. Fewer Sonic games were released than Mario games, and they were, at least in the 2D era, confined to only two consoles. Sega's MegaDrive (Genesis) was where he got his start, but in a move unusual in the games industry, almost all of the *Sonic* games were ported backward to Sega's Master System console, which was a hardware generation older than the MegaDrive. That's not surprising when you consider Sega has a massive following in Brazil where, as late as 2003, the Master System saw new releases.

These Sonic sprites show a clear progression, and it's easy to see how Sega refined him. Sonic was initially a less vivid blue. This was changed in *Sonic 2* to help him stand out against the background. He stayed that way for the rest of the series.

Running Sonic
There was a steady progression in Sonic's limbs. The first Sonic had very skinny arms and legs, but by *Sonic 3* they were quite a bit more muscular in appearance. Sonic's shoes, perversely, seemed to get cleaner as time went on—by the third game they were gleaming, and they seemed to be much rounder. If you look carefully, you can see how Sonic's running skills have straightened out his feet too. The feet on the first-generation Sonic tended to orient the soles inward. Hedgehogs are known for their unusual gait, but this might have been too unusual for Sega: Sonic's feet were snapped into place from the second game onward.

Standing Sonic
The standing position saw fewer changes than the running one, at least between the first and second games. Except for some widening

1. *Sonic the Hedgehog* running sprites
2. *Sonic the Hedgehog* standing sprites

1.

2.

Genre/pixel histories

Despite starting on the 16-bit Genesis, Sonic saw more releases for the 8-bit GameGear and Master System platforms. Sonic always looked pretty good in 2D, but the 3D versions were, except for the MarbleMadness-alike arcade game, completely unattractive.

of the eyes in Sonic 2, giving Sonic a less angry appearance, the sprites are identical. The adjustment from the second to third games is much more extreme. Sonic was given much larger hands, and because of their placement Sonic looks almost chubby. Previous iterations had him thrusting his chest forward with his arms back, but now he appears more relaxed. His cheeks are also rounder in appearance, and there's now the suggestion of a mouth, where there never was one before.

Master System / GameGear port

It's unusual for a manufacturer to support two different generations of hardware at the same time. If this does occur, it doesn't happen for long. Unusually, Sega was able to maintain continued success with two generations of hardware. Sega's portable color game system, the GameGear, was a modified Master System internally. It added an enhanced color palette (4,096 colors instead of 64) but used a screen with a lower resolution. Except for the physical size of the cartridge, Master System games would run without modification on the GameGear, so Sega was able to provide games for several markets with each game it produced. Because of its success in Brazil, and to a lesser extent in Europe, the Master System saw releases of several Sonic games long after support for the system would normally have been discontinued. By programming for the GameGear first, Sega could supply the GameGear audience with new games and then release these same games for the Master System in other countries.

It wasn't always a perfect process. Since the Master System had a much higher resolution than the GameGear, games tended to run much slower on the Master System. The CPU, taxed to the limit running the smaller GameGear games, was overloaded when it was expected to handle the higher resolution.

1. *Sonic the Hedgehog* (Sega Genesis 16-bit)
2. *Sonic the Hedgehog* (Sega Master System 8-bit)

1.

2.

Sonic the Hedgehog sprite history

The conversion process from 16-bits to 8-bits was very unusual. Normally a character is moved to newer, better platforms. Sonic was consistently being moved backwards to the Master System.

Despite the sizeable market for these games, the aging hardware was not capable of accurately replicating games from the MegaDrive. Fewer colors and less powerful hardware demanded that sacrifices be made, with none so evident as the reductions made to the Sonic sprite itself.

The first two Sonics were identical except for the color of Sonic's shoes. Significant changes were made to the third Sonic, making him cuter and rounder, like the MegaDrive *Sonic 3* sprite. Sega was also unusual in being able to cross genres with its mascot. Normally, a flagship product isn't changed very much to avoid alienating fans, but when players raved about the pinball levels in the original Sonic games, Sega listened. *Sonic Spinball* was the result, a kind of adventure pinball where Sonic spent most of the game rolled up into a ball bouncing around a giant pinball machine. Pinballs aren't very big,

and Sonic had to be reduced in size once again. The GameGear, with its lower price point and library of short "throwaway" games received several updates that the MegaDrive didn't. Sega, it seemed, was more willing to experiment on the GameGear since the audience was smaller, the costs were lower and the risks—of flooding the market or diluting the value of their mascot—were reduced. These games included new levels and game types as well as several additional revisions of the Sonic sprite. While the hardware wasn't capable of displaying the large, detailed sprites seen on the MegaDrive, Sega's artists still managed to achieve some very acceptable results.

Like many game characters, there was a time when Sonic went through a very ugly stage. Sega's effort to keep the franchise new and fresh went completely awry when the developers tried to make *Sonic 3D*. Using pre-rendered graphics Sega

put Sonic in a 3D world, which not only made him lose the majority of the charm he possessed, but also made for very poor gaming. The 3D Sonic was, like most pre-rendered sprites, blotchy and hard to recognize. Note the use of stippling to simulate a shadow under the 3D sonic. This was a common technique on older systems that weren't able to create real transparency effects. On an old television these black pixels would blur a little against the background, providing a pseudo-transparent shadow.

What is especially surprising about these 3D Sonic sprites is how Sega also ported this 3D version to the Master System. There was no way even the talented artists at Sega could make a pre-rendered Sonic sprite look good in 16 colors.

The GameBoy Advance saw the first major change to the 2D Sonic Sprite. Not counting the 3D Sonic games, which are best considered

1. *Sonic the Hedgehog* running sprites
2. *Sonic the Hedgehog* standing sprites

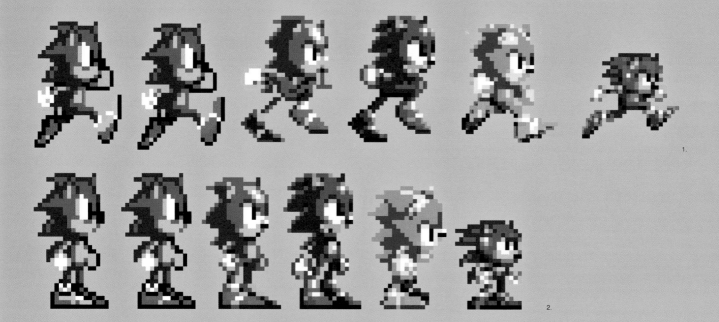

1.

2.

Genre/pixel histories

Some games re-use sprites from one game to the next, but Sonic, like Mario, was given the royal treatment and was redrawn for almost every new game.

aberrations, the GameBoy Advance Sonic sprites are the only deviation from the original sprite. Whereas all the changes made before were tweaks and small changes, the GameBoy Advance saw a new Sonic that was much more like the promotional art Sega had always offered, and less like the old sprites. Taller, thinner, angrier, this Sonic is edgy and very much more "in your face." This Sonic might best be considered "x-treme," a Sonic for the new generation.

1. *Sonic the Hedgehog* standing sprites
2. *Sonic the Hedgehog* running sprites
3. *Sonic the Hedgehog* (MegaDrive)
4. *Sonic the Hedgehog 3D Blast* (MegaDrive)

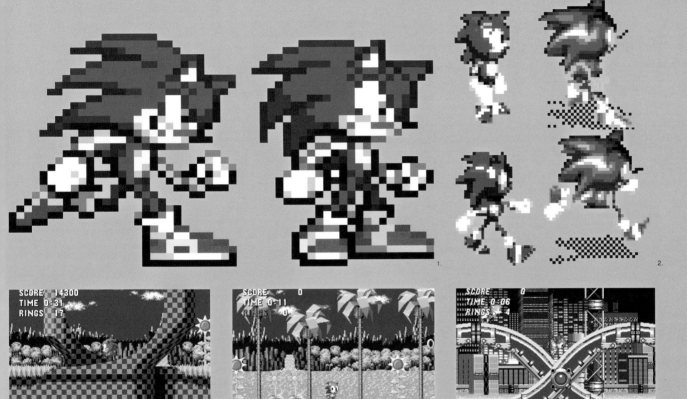

Sonic the Hedgehog sprite history

The 3D Sonics (top) were terrible. The GameBoy Advance Sonics were the first to be reused, unchanged, from one game to another.

171

Bonk sprite history

Less well-known to non-gamers, perhaps, Bonk was a sprite that captured the imagination and wallets alike. Although visually less striking than his peers, his design mixed simple lines with more complex gameplay.

Back in 1989 NEC needed a hit game. The PC Engine console, a runaway success in Japan, was losing momentum in North America, where as the TurboGrafx-16, it was steadily losing ground to Sega's Genesis (MegaDrive). Technologically superior in many ways to all of its competition, the TurboGrafx offered more and brighter colors to artists. Hudson, a software company with a fledgling chip division, created the chips that powered the machine and it was to them that NEC turned for a big new game. Hudson didn't disappoint.

Bonk's Adventure was released in 1990, to critical acclaim. A platform game like no other, it featured a caveman protagonist who attacked his enemies by head-butting them, or by leaping into the air, spinning, and landing on them head first. It was a radical departure from the staid old Mario routine of runing and jumping feet first. It was

immediately a top seller for NEC, one of its first big hits, and Bonk quickly became the unofficial mascot for the system.

Graphically, it wasn't particularly noteworthy, though it offered impressively smooth animation and played very well. It was the variety of animations that impressed—it was packed with unique new moves and the graphics to go with them. To climb walls, Bonk would latch on with his teeth and gnaw his way upward as fast as the player could hit the button. He was also fond of meat, and whenever he'd find some he'd greedily eat it, launch himself into the air and, for a short time, become more powerful. By the time a player reached the first boss he'd already run, jumped, climbed, become invincible, crossed the spiny back of a dinosaur, crawled inside the dinosaur's mouth, and worked his way out the other end.

1. *Bonk* sprites (GameBoy)
2. *Bonk 3* sprite (PC Engine)
3. *Bonk 1* sprite (SNES)

1.   2.   3.

Genre/pixel histories

Bonk was always an underappreciated success. The first *Bonk's Adventure* game was the flagship title for the TurboGrafx 16, but the masses were not as interested in headbutting meat-eating cavemen as they were in block-smashing plumbers.

It was a refreshing change of pace. When the first sequel came out in 1991, it didn't stray too far from the successful formula. There were plenty of new bonus stages, and some new moves were added, like the ability to climb waterfalls. The big change was in Bonk's appearance. He had become softer, rounder, a little more cuddly—no doubt in an effort to improve his marketability. The second game begged for a sequel.

The third game offered yet more new features, including two-player modes and the ability to become very tiny or grow to enormous proportions. In spite of these additional features the game wasn't as big a success. The time had come and gone for the PC Engine, and this final Bonk was too late to sell in significant numbers.

However, the game and character design were so successful Hudson couldn't let them die. Several other systems received ports, and in 1993 and 1994 Nintendo's NES and SNES each saw a version using the same character sprites, though with fewer and more colors, respectively.

It's never easy converting sprites from one system to another. Though the GameBoy sprites are scaled-down versions of the first two PC Engine games, they managed to capture the spirit of Bonk nicely, and in only two colors.

1. *Bonk* sprites (PC)
2. *Bonk* sprites (SNES)
3. *Bonk* sprites (Arcade)
4. *Bonk* (NES)

1.

2.   3.

FREAKTHOROPUS COMPUTERUS

PUSH START BUTTON

©1993 HUDSON SOFT
©1989 1993 RED

4.

Bonk sprite history

Hudson kept trying, and Bonk was released for the Famicom, GameBoy, Super Nintendo, and with a little help from Kaneko, the arcade as well.

The original Bonk was a little rough, with a crooked mouth, large ears, and a tuft of hair on the back of the head. The first GameBoy sprite is identical in design. While functional, it didn't take Hudson long to make improvements. The second-generation Bonk is much rounder, with a smaller ear and no more hair.

The first SNES version used so many colors that the resulting Bonk sprite looks almost disturbingly realistic, as if the artist had digitized a real head and pixeled in new features. New sprites were used for the second SNES version, though curiously they were now much smaller in size, almost a cross between the second-generation

Bonk and the GameBoy version. This design persisted until the very last, when Hudson went too far. The second SNES version, second last on the column below, used a less attractive, smaller sprite. This was to be the last console release of a Bonk game for nearly a decade.

The last *Bonk* in this evolution was released in 1994, the same year as the final SNES game, and was the only one not to be created by Hudson. Released in the arcade by Kaneko, this was an entirely new game. The graphics, like most arcade games of the time, were vivid and excessive, with more moving objects in one short level than in any three stages in the console games.

It has a unique style—a *Bonk* that could almost be a hybrid of the first two generation sprites but for the fact it was released four years later.

The second game, *Bonk's Revenge*, started a strange trend of having Bonk, now more natural in appearance most of the time, become more bizarre-looking when he ate meat. In the original game, scars would appear and Bonk would become darker, but now the effect was even more striking and Bonk appeared mutated and lumpy. In an attempt to make his meat-addled appearance more striking even than this, Hudson ran headlong against the wall of good taste—and then pushed with all its might.

1. *Bonk* sprites (PC Engine)
2. *Bonk* sprites (SNES)
3. *Bonk* sprites (SNES)

Bonk sprite history

When Bonk saw his final release on the SNES, it was clear Hudson was grasping at straws. Instead of a simple and fun core game, Bonk had more than a half-dozen new shapes to transform into, and the game became tedious—though very good looking.

Bonk enjoyed a half decade of success before Hudson went too far. The Super NES versions of Bonk went over the top, straying too far from the simplicity of the original. In the last SNES release Bonk could transform into a wild variety of creatures with new skills, depending on the color of meat he ate. In their rush to stuff as many transformations into as small a space as they could, the designers forgot to let the player enjoy them. The series collapsed and was nearly forgotten until 2003, when Hudson re-released the first game, using polygons instead of sprites, on the GameCube and PlayStation 2.

1. *Bonk's Adventure* (NES)
2. *Bonks Revenge* (GameBoy)
3. *Bonk* (Arcade)

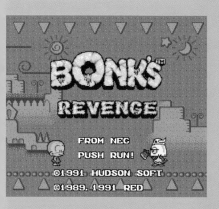

5.3:
Bonk sprite history

The first PC Engine game (top) was rather plain compared to its sequel (center), but neither compared to the visual extravaganza of the arcade game.

175

Arthur Ghouls 'n' Ghosts sprite history

Three games and thirteen versions have seen this character design become established as a firm favorite

Ghouls 'n' Ghosts is one of Capcom's big success stories. Thirteen versions of the series' three games have sold over four million copies since 1985. The series started in the arcades and has been ported to the NES, SNES, GameBoy, GameBoy Advance, NeoGeo Pocket Color, Playstation, Saturn, Wonderswan, PC Engine, MegaDrive, and recently to NTT DoCoMo's i-mode devices. It features some of the very best art Capcom produced, and the varied platforms, each with different capabilities, meant there were lots of adjustments for its artists to make.

The first arcade game set the stage with remarkable capability, considering the hardware abilities of the time. The main character, Arthur, is an armored knight who attempts to rid the world of ghosts (and possibly goblins) and save his lady from their evil clutches. His

journey is, in several ways, an arduous one. Throughout his adventures in 13 releases Arthur himself has been stretched, squashed, rendered colorless, and redrawn many times.

Arthur's first incarnation is in the arcade. He's 19 pixels wide and 31 tall, with a blue eye and a dashing scarlet belt that makes the ladies wild. By the time he made it to the small screen on Nintendo's NES he'd nearly lost it all. Almost completely colorless now, there's no sign of the blue eye, and he's lost three vertical and five horizontal pixels. The average arcade screen used about 56 colors, but the poor old NES could manage only 12.

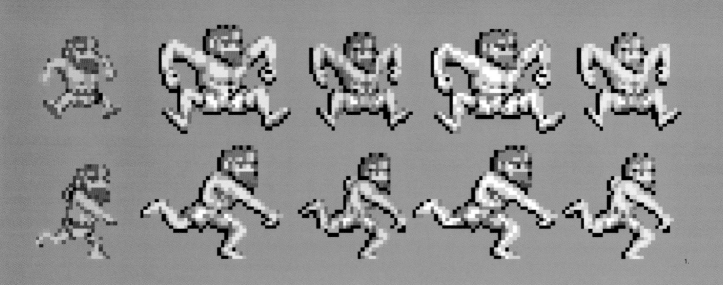

1.

Capcom was no stranger to re-using sprites. Many of their games used the same images from one game to another, and Arthur was no exception. For a long time, except for size changes (and the resulting tweaks) the only significant change was from the first *Ghosts 'n' Goblins* game to its sequel, *Ghouls 'n' Ghosts*.

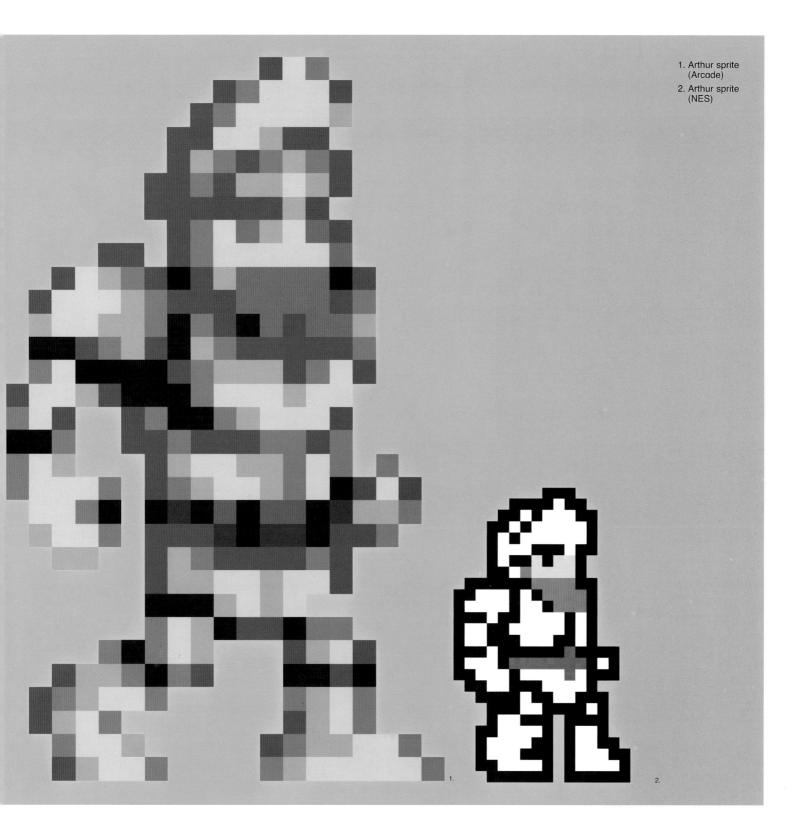

1.

2.

Arthur Ghouls 'n'
Ghosts sprite
history

Sometimes there
were extreme
differences from
one platform to
another, as this
arcade and NES
Arthur show.

177

The second game saw a much more refined Arthur. A little older, a little taller—up seven pixels now to 38. Interestingly, the first game in the arcade used some hardware trickery to stretch the entire screen by nearly 50 percent, so Arthur would have appeared nearly the same width in both games. The sprite has been completely redrawn with a much simpler look. The original had 12 colors, but now 13 are used. Arthur didn't fare so well when it came to the consoles, however. While the 6-megabit MegaDrive (Genesis) cartridge was considered huge at the time, it was less than half the size of the arcade version, and Arthur was reduced to 9 colors. The 8-megabit SuperGrafx version gave Arthur a more detailed world to quest in, but cut his palette down to

a mere six colors. The SuperGrafx Arthur also lost one vertical pixel; however, unlike the MegaDrive version, he kept both hands.

Arthur's good looks took a beating when the Atari ST version of the game was released. The ST's paltry 16-color palette didn't do justice to many games, and the underpowered CPU demanded that programmers cut the screen size by a third. Arthur really suffered with only five colors dedicated to his sprite, and he was reduced in size to 18 x 31 pixels— one pixel smaller than he was in the first game.

Super Ghouls 'n' Ghosts
Super Ghouls 'n' Ghosts was the third game in the series, and except for a GameBoy Advance remix, it

was the last game in the series. The Super NES, still new to developers when *Super Ghouls 'n' Ghosts* was released, was unable to draw levels as large as the second arcade game, so instead had the same resolution as the first game. When displayed on a TV the image was, like the first arcade game, stretched horizontally by nearly 50 percent. The original 23 x 37 sprite was, when stretched, almost square. In the image below you can see the progression, with Arthur badly stretched to mimic the effect of playing on a real monitor.

Ghost Trick
The sprites, opposite, are Arthur reimagined for some other uses. The left image is taken from *Ghost Trick*, a single-screen minigame

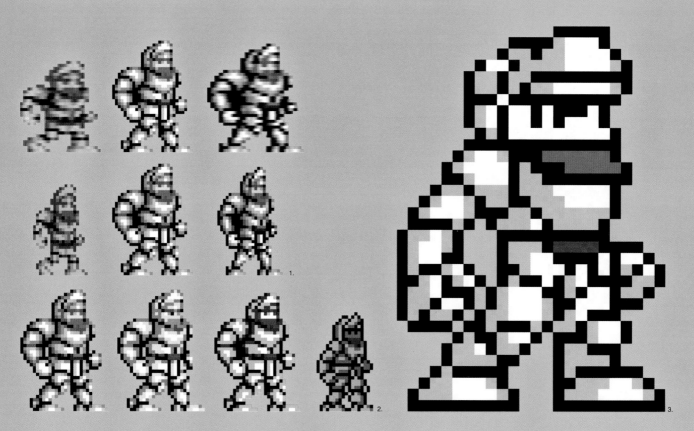

1.

2.

3.

Genre/pixel histories

In most cases Arthur was copied directly from one platform to another. Sometimes a skinny Arthur was stretched out to fill the screen a little better, but on rare occasions, like the NeoGeo Pocket game *Millennium Fighters*, he was completely redrawn.

from SNK's NeoGeo Pocket Color game, *Match of the Millennium*. His short little friend is from the Wonderswan *Ghouls 'n' Ghosts*, a Japan-only release on the Wonderswan portable. Surprisingly, the Wonderswan version has much better animation than the three shades of gray seen below might imply, and Arthur is smoother than in any other game in the series.

You can see how, given the limits of the target hardware, artists struggled to capture Arthur's spirit. When it's not possible to reuse the same sprite, designers are forced to

make changes and sometimes these changes don't turn out too badly. The tall Arthur is a bit of a surprise. He's not a playable character, but rather a kind of backup, a special attack in Capcom's *Marvel vs Capcom*. It makes you wonder what a pixel-Arthur would look like on a modern platform, fighting ghouls, ghosts, and goblins in a high-definition style.

1. Arthur sprite, *Ghost Trick*
2. Arthur sprite, Wonderswan release

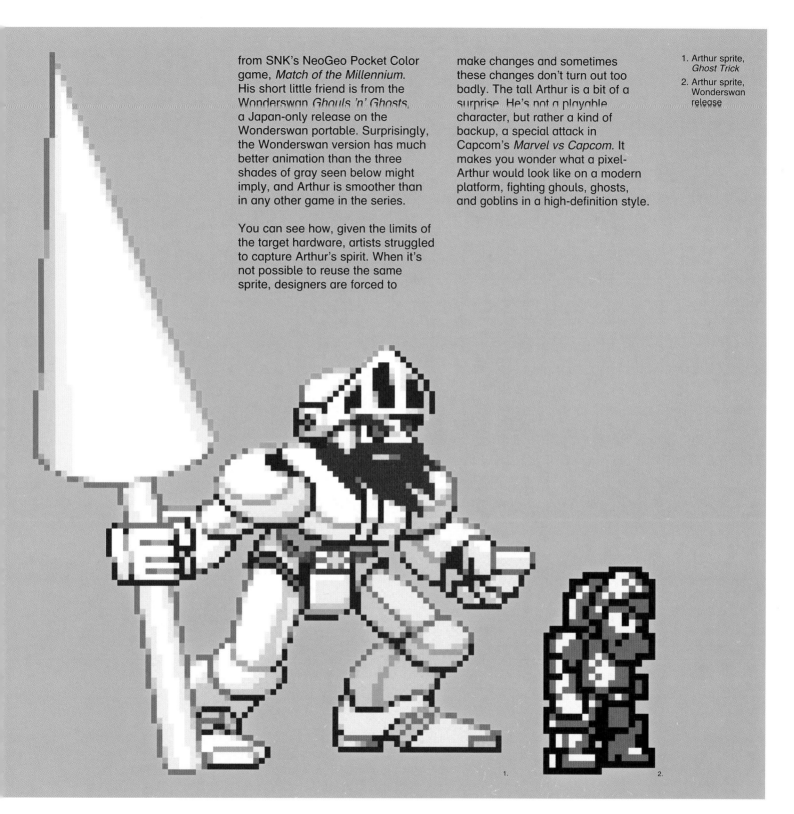

1.

2.

Arthur Ghouls 'n Ghosts sprite history

Arthur did see significant changes on some obscure platforms. The arcade *Marvel vs Capcom*, where he played a cameo role, and the WonderSwan *Ghouls 'n' Ghosts*, were about as wildly different as could be imagined.

179

Castlevania sprite history

Konami's title is massively popular in some places, but almost unknown in others. How have the character designs developed over 10 gaming systems?

Konami's *Castlevania* is an unusual game. It's incredibly popular in the Western world, but is a consistently poor seller in its native Japan. Despite this, it is still a mainstay of Konami's game lineup, and has so far graced more than 10 different game systems. Except for a few unpopular attempts at 3D, the series is still 2D.

For most of its run, the *Castlevania* protagonist has used a whip as the main weapon. Only a few games have strayed from this, but even games that offered a new weapon had a whip that could be equipped, or even a whip-wielding alternate character. The series started on the Nintendo Entertainment System hardware, and has appeared on most systems since then. The only Nintendo platform since the NES without a *Castlevania* was the Virtual Boy.

The main character frequently changed, and in some instances a female took the lead. The artists at Konami were very successful in creating characters from one hardware generation to the next, maintaining the same spirit from the original games while taking advantage of the new hardware's capabilities.

1. *Castlevania* sprite (NES)
2. *Castlevania 2: Simon's Quest* sprite (NES)
3. *Castlevania 3: Dracula's Curse* sprite (NES)

Genre/pixel histories

These three NES sprites are crude because of the hardware limitations, but it seems clear that Konami had a solid design in mind from the start.

1. *Castlevania* (PlayStation)
2. *Castlevania* (PlayStation)
3. *Castlevania Adventure* sprite (GameBoy)
4. *Castlevania 2 Belmont's Revenge* sprite (GameBoy)
5. *Castlevania Legends* sprite (GameBoy)

1.

2.

3.

4.

5.

Castlevania sprite history

Few images can contrast the differences in a sprite's evolution than these. The PlayStation screenshots are simply light years beyond the pale GameBoy sprites.

181

During the NES's reign, the hero changed very little. Only a handful of pixels in the shoulders changed from one sprite to the next. It could be said that from the start Konami had maxed out the hardware, using all four available colors, and gaining only a few pixels in size by the time the third game was released. The original GameBoy presented serious challenges for all pixel artists, but Konami's artists rose to the task. The heroes in all three GameBoy games retained their whip-toting appearance, even while inexplicably becoming left-handed.

This first image is from the only arcade *Castlevania*, a 1988 release called *Haunted Castle*—a truly awful game even diehard fans of the series can't stand. Konami suddenly figured out what to do with new, modern hardware, and in 1993 the X68000, a super-powerful Japanese computer,

saw an upgrade in visuals better than the arcade version of a few years earlier. Next came the PC Engine version, which would become the gold standard for many years. The same sprite was used in both the second Super NES game and the Playstation and Saturn sequels. In 1994 Konami seems to have fallen from its perch: the MegaDrive version is inexcusably as garish as the arcade version was ugly.

For a while Konami pulled out all the stops with the PlayStation and Saturn versions, but went back to basics with the GameBoy Advance.

The next major change in the graphic appearance was with the launch of the GameBoy Advance. *Castlevania: Circle of the Moon* was released alongside the system, and it was immediately slammed in the gaming press for being too dark to

play on the GBA's less than vivid screen. The next GBA release featured a larger, though strangely less detailed, protagonist. As if realizing the mistake, Konami made a large and detailed hero for the final GBA outing, but again forgot to give him a whip.

The last *Castlevania* game to be released so far is the top-selling *Castlevania: Dawn of Sorrow* for Nintendo's DS. It sports some of the very best pixel art ever seen, and may well be the standard bearer for aspiring pixel artists for the forseeable future.

1. *Haunted Castle*, (arcade)

2. *Akumajo Dracula*, (X68000)

3. *Dracula X: Rondo of Blood*, (PC Engine)

4. *Castlevania: Bloodlines*, (MegaDrive)

1.

2.

3.

4.

Genre/pixel histories

The arcade Castlevania was, without question, ugly. It's a lucky thing another game was made at all. After experimenting a little with the X68000 and Genesis versions, Konami settled on a Richter for a while. The PC Engine version was re-used in several sequels.

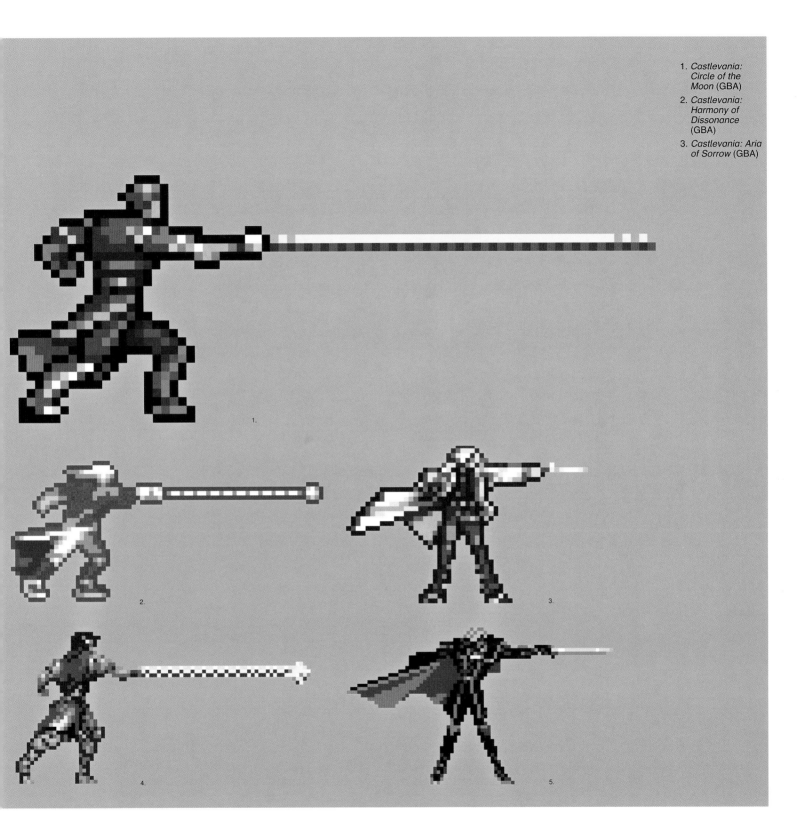

1. *Castlevania: Circle of the Moon* (GBA)

2. *Castlevania: Harmony of Dissonance* (GBA)

3. *Castlevania: Aria of Sorrow* (GBA)

1.

2.

3.

4.

5.

Castlevania sprite history

Konami went with a very different look for all three GameBoy Advance and DS games.

183

Glossary

2D
Two dimensions, flat, without depth. Most games before 1996 were 2D. A defining aspect of 2D games is the art, most often handcrafted one pixel at a time, and often more vibrant than 3D games can be. There is a large and vocal subset of game players to whom the days of 2D games are a golden age, where games were "pure and fun." The days of 2D gaming were without doubt the apex of pixel art, but this was largely a result of underpowered hardware. Earlier computers couldn't handle 3D animation, so designers pushed the limits to try and make more attractive images within these two dimensions.

3D
Three dimensions, not flat. The move to 3D games has to a large extent killed off pixel art, relegating it to portable platforms almost exclusively. However, it has created a new wave in realism-based games. Modern games are constantly pushing the boundaries of the possible, creating realistic worlds out of polygons and raw CPU power. Creating games from 3D requires an incredible increase in computing power to realize the new, hyper-detailed worlds. It comes at a cost, however—many older players blame the rise of 3D games and a new push for realism over style, and accuracy over creativity, for ruining games.

Aliasing
When the pixels at the edge of a polygon or sprite are significantly brighter or darker than neighboring pixels they can become very visible, and can distract the eye of the viewer. This is called aliasing. The effect is also known as "jaggies." When the PlayStation 2 was new developers had trouble eliminating the aliased effect, and early games were known for being flickery and jagged.

Anti-aliasing
A technique that reduces the visual impact of high- contrast adjacent pixels by blurring them with their neighbors. While this technique inevitably results in some blurring of the image it is widely considered preferable to the jagged, distracting alternative. Older hardware could not anti-alias "on the fly," and instead relied on the artist's skill to do this blending when the image was created. Modern hardware typically performs this function automatically, primarily to reduce jagged, aliased edges on polygons.

Aspect ratio
The relationship of a display's width and height, typically denoted as x:y. Normal television screens have an aspect ratio of 4:3, or four units wide and three units tall. Widescreen TVs use an aspect ratio of 16:9. Portable game systems don't adhere to any specific aspect ratio.

Bit
Binary digit. In a digital system, such as a computer or games system, all data is processed as a series of electrical charges, which are then stored in binary code, with each binary digit, a one or a zero, representing a positive or negative charge. The longer the binary "word," the more data can be stored in it—hence 8-bit, 16-bit, and so on.

CPU
Central Processing Unit. The part of a computer that typically does the majority of processing. The CPU is a general-purpose processor. It is often, but not always, assisted by one or more task-specific coprocessors, which relieve the CPU of some specialized burdens. Most game systems and modern computers have a graphics processor, and often several extra units to handle input and output, audio, memory access, and other tasks.

CRT
Cathode Ray Tube. Before LCD and plasma screens the CRT display was, for all intents and purposes, the only system available. Images are created by sweeping a spray of electrons across a screen coated with phosphor dots. By varying the strength of the beam, different brightnesses can be

shown, creating a grayscale image. The use of different colored phosphors on the screen creates the illusion of a full-color image.

Digitized graphics
Graphics converted to a digital form from real-world objects are called digitized graphics. While used in games as early as 1983 (Bally/Midway's *Journey*) it wasn't until *Mortal Kombat*, another Midway game, that they really hit the mainstream. Underpowered hardware typically resulted in compromises in animation and color that crippled the effectiveness of this graphic technique. By the time hardware was able to represent digitized images effectively the gaming world had moved on to polygons. There are no games featuring digitized graphics to which one could point and say, "That's how it should be done." It's just not an attractive technique.

GIF
Graphics Interchange Format, a graphics file format created by the company Compuserve. It is a lossless format that perfectly preserves an image. However, it is limited to a maximum of 256 colors. It is best used for computer-generated images, and fares poorly when used for photographs or other digitized content. A company called Uniserve claimed patent rights over the compression used in GIFs, and long after GIF was a dominant file format companies were suddenly asked to pay royalties. It was because of this shakedown that the PNG format was created.

GPU
Graphics Processing Unit. The GPU, typically a separate chip from the CPU, handles the tasks of drawing images on the screen, freeing up the CPU for other things. In early consoles, the GPU was a less capable slave chip, designed to do only a few tasks but do them quickly and efficiently. As game hardware advanced, the GPU became more powerful, and modern video chips rival the CPU in terms of power and flexibility.

Hard scaling
When an image is increased or decreased in size it is "scaled," and several techniques facilitate this. Hard scaling is the fastest way to achieve the new size, however it is typically also the ugliest. When increasing an image from 100 pixels to 110 pixels, hard scaling simply duplicates every tenth pixel. This requires very little CPU power, but the results are typically jagged, uneven, and very unattractive. Hard scaling fares a little better when an image is reduced in size. However, the new, smaller image may be missing important pixels—the pupil of an eye for example. (See Soft Scaling)

LCD
Liquid Crystal Display. LCD screens are fixed-resolution, flat-panel, raster displays that offer geometrically precise images. They're lightweight, long-lasting, and, increasingly, affordable. A relatively new technology, LCD screens have been used in games from the very earliest days, and remain the preferred technology for portable gaming. LCDs comprise a layer of liquid crystals between two sheets of glass. An electric current is applied to small regions of the screen, preventing light from passing through, creating areas of light and shadow and, ultimately, an image.

LED
Light Emitting Diode. A small electronic device that emits light. LEDs are energy efficient and long lasting, however until very recently, they were only available in a limited number of colors. Once used for rudimentary portable gaming, LEDs are found in nearly every electrical consumer device sold today. Ultra-bright LEDs are used for traffic lights, train signaling, and lights in automobiles.

Lossless compression
If a digital data can be compressed (to minimize storage needs or to speed its transfer over a network) and then recreated identically to the original, the compression method is considered lossless. GIF and PNG are two lossless image formats.

Lossy compression
In order to achieve the most efficient compression some data is discarded, resulting in a digital entity that is of lesser quality than the original. Typically perceptual encoding is used, where the compression scheme discards data that humans can't normally distinguish. In images, blue colors may be stored at a greatly reduced resolution due to the human eye's inability to see as much blue detail as green. JPG and MP3 are two lossy data formats.

NES
Nintendo Entertainment System, the 8-bit game system from Nintendo that put the company on the top of the video game heap, and resurrected video games from the historical dustbin after the mismanagement of Atari dumped them there.

PDA
Personal Digital Assistant. A usually business-oriented portable computer, typically handheld, and with a touch-sensitive screen. Newer units sport full VGA (640x480) resolution and over 250,000 colors. While the screen quality is usually exceptional, the lack of a proper game pad makes the platform something less than suitable for gaming.

Pixel
Pixel is a contraction of Picture Element, and is the smallest component of an image. With the exception of older and specialized vector hardware, every game and computer system creates images from a massive array of pixels.

PNG
PNG is a lossless image format similar to GIF, with support for a full 24-bit color palette, more efficient (and royalty free) compression, and true 256-level alpha channel transparency. Unfortunately PNG has not dethroned GIF as the dominant lossless image format, due to patchy support within the Internet Explorer browser.

Glossary

Polygon
Polygons are the primary elements of a computer-generated 3D image. Each polygon has three sides, joined at the edges to other polygons, creating a much larger whole. After the polygons are assembled, textured, and illuminated they can be very lifelike and the polygons themselves are usually invisible.

Prerendered
When creating graphics for a game, an artist might start with a 3D object, which can be posed, lit, and flattened into a 2D sprite for use on less powerful hardware. The result is, at best, a moving sprite that looks like a detailed polygonal object. Typically, the compression results in an image which is, sadly, not nearly as pretty as the original. *Donkey Kong Country* is perhaps the best-known game to use prerendered images.

PSP
PlayStation Portable, Sony's attempt to bring portable 3D gaming to the masses.

Raster
Raster typically refers to an image made of pixels. Unlike a vector image, raster images cannot be scaled cleanly. The number of pixels in a raster image is a hard quantity, and cannot be adjusted cleanly.

Raster-scan monitor
A raster monitor is a CRT, like older TVs or most arcade monitors. Images are drawn by an electron beam that sweeps from side to side, stacking layers of an image from the top of the screen to the bottom. Raster CRT monitors can vary the density of these lines (called scan lines) so that a pixel can be stretched slightly. By tuning a raster CRT a pixel image can be stretched or reduced in size without any appreciable loss of quality or clarity, which is one of the few advantages offered by this nearly century-old technology over plasma and LCD screens.

Resampling
Resampling is the act of taking a digitized entity and changing the rate of sampling. In the same way that enlarging an image cannot magically insert missing pixels, "upsampling" an image does not make a digitized entity better quality, so typically this is a one-way process. When resampling a signal to a lower rate the quality is invariably reduced, so wherever possible the original analog signal or image should be used as a source.

Resolution
The number of pixels arrayed horizontally and vertically on a screen is the screen's resolution. The resolution of a video game screen is listed in pixels. For example most Super Nintendo games had a 256x223 resolution. The resolution of a GameBoy was 160x144.

Rotoscoping
Before digitized graphics there were rotoscoped graphics, and graphics animators used the technique to create accurate motion in cartoons. In games, the lack of sophisticated digital cameras or digitizing hardware required alternate techniques. Rotoscoping was, in essence, tracing. A frame from a filmed sequence was displayed and the artist would "pixel" over the original image, then advance to the next frame and do it again. One of the first, if not the very first games to use rotoscoping was Jordan Mechner's *Karateka*, and later *Prince of Persia*.

Sampling
Sampling is core to the digitization process. When an analog signal is being digitized it will be sampled. Sampling is, in essence, the evaluation of the analog signal at a specific frequency. The accuracy of the digital recreation can be determined, in part, by the sampling frequency—the more often it is sampled the more accurate it will be. For example, CD audio is sampled 44,100 times per second, and is typically considered "accurate enough" at that speed. High sampling rates require more space and CPU

power to manipulate, as higher sampling rates create a digital entity with more data.

Saturation

Saturation refers to the amount, or vibrancy, of a color. A highly saturated image will appear to have very bright colors, where low saturation will appear grayer. An image with zero saturation would be entirely grayscale, without any color at all.

Soft scaling

When an image is increased or decreased in size it is "scaled", and several techniques will facilitate this scaling. Soft scaling is a CPU intensive way to achieve the new size, however it is typically unattractive. When scaling an image up or down it is resampled, so that every new pixel is, where necessary, blended with its neighbors to produce an approximation of the original image. The resultant image suffers many artifacts. Blurring is the first problem, because of the pixels blending with their neighbors. A noticeable decrease in contrast results as well, as pixels that were much brighter or darker than their neighbors are combined.

SNK

Shin Nihon Kikaku ("New Japan Product") is a Japanese game manufacturer that shot to fame on the backs of two products: the wildly successful NeoGeo arcade and home hardware, and SNK's fighting games. While not a minor player before the NeoGeo, SNK's games never reached blockbuster popularity. After the company launched the NeoGeo, along with a steady stream of progressively more innovative fighting games, the company became a world leader. However, the success was short lived, and, like many companies, it suffered mightily when the popularity of game arcades started to fade. The portable NeoGeo Pocket was, at the time, much better than the GameBoy Color, but SNK didn't have the resources to market it successfully. Best known for the *Samurai Shodown*, *King of Fighters*,

and *Fatal Fury* series, SNK is now relegated to small-time status, porting moderately successful retro hits to the PS2 and mobile phones.

Sprite

A moving image on a computer screen. In the early days of games, computer hardware was very general-purpose, and graphics were all drawn and manipulated by the CPU. Graphics at this time were not handled in a specific way, and no distinction was made between drawing background images, characters, or overlays. Along with dedicated video hardware, however, came new capabilities. Specific tasks were handled by the GPU, such as moving a background image, and programming became much easier as the hardware reduced more and more of the programmer's responsibilities. Sprites—individual movable objects, were now separated from the background in hardware instead of software. At first the limited speed of the hardware restricted the use of sprites—only a small number could appear on one screen, and fewer still on the same horizontal line. Newer, faster hardware added the ability to manipulate more and more sprites, until eventually there was no real limit.

Vector

In a geometrical sense, a vector is a combined evaluation of movement, of direction and speed. In games there are two kinds of vectors, both very similar in origin but not in the way they're displayed. In the early days of arcade games specialized monitors were used that would draw lines in any direction, unlike raster monitors which draw them in a stacked horizontal pattern. To draw a box a line would simply be drawn from one corner to another, until the shape was completed. Games like *Asteroids* and *Star Wars* used vectors to create the game screen, with each game object made from bright lines on a black screen. In more modern games vectors are very similar, but are rendered on a raster monitor. In this case the lines are drawn in a much more CPU-intensive and flexible

manner: shapes can be filled with colors or patters, lines can be different colors or widths, and so on. Both kinds of vector drawings, consisting of a numerical description instead of a pile of pixels like raster images, can be scaled smoothly without losing detail.

Vector monitor

A vector monitor uses the same CRT as a raster monitor, but utilizes completely different electronics to drive the electron beams. Instead of drawing an image by sweeping the beam left to right and top to bottom, a vector monitor draws a line from one arbitrary point on the monitor to another. Vector monitors were considerably slower in drawing speed than a similar raster monitor, so shapes could not be filled. Wireframe graphics and line drawings were, while primitive, very flexible. Unfortunately the limitations of the format, combined with the rapid increase of capability of less esoteric raster game hardware, combined to limit the vector monitor's successful time to a mere half decade.

Index

A

3D graphics
    current generation devices 32–3
    developers 84, 86, 93, 100, 113
    motion blurring 53
    pixels 56, 86, 100
    platform games 156–8, 170
    pre-rendering 56, 144, 156,
        158, 170
    retro gaming 84
    role-playing games 143
    see also polygons
Adobe Photoshop 64, 66, 82, 99–101
Adventurevision 14–15
Amiga 2000 72
anti-aliasing 8, 63, 99, 106
Apple II 132–3, 138
arcade games 148–53, 164–7, 173–5
Army of Trolls 64–7
ASCII characters 132
aspect ratios 122–3, 125, 178
Atari 2600 8
Atari Jaguar 157
Atari Lynx 28–9
Atari ST 178
Autodesk Animator 59

B

backlighting 51
Barnard, Charles 79, 82–4, 86
Blue Label Games 87–97
blurring pixels 52–3, 126
BMP format 90
Bonk's Adventure 155, 172–5
Burgertime 154

C

Cabybara 98–103
Capcom Play System One
    (CPS-1) 150–3
Castlevania 157, 158, 180–3

cathode ray tubes (CRTs) 12–13,
    50–1, 53
cellphones 44–9
    Cabybara 98–103
    converting graphics 118–21
    current generation 32
    developers 78–131
    file formats 90–1
    history 6, 19
    memory usage 80, 82, 90–2, 96
    Nieborg, Henk 62–3
    porting 94, 118–31
    production cycle 94–5
    resolution 92–3
    role-playing games 145
    sprite formats 88–9
Chan, Anthony 98–103
character development 98–9, 104–7,
    139–42, 153, 160–83
Chrono Trigger 140, 143–4
clones 56, 80
color displays 42–3, 48–51
Commodore 64 60
contrast 48, 51
converting graphics *see* porting
Cook, Chris 80
Corel Graphics Suite 74
Corel PhotoPaint 105, 106–7, 109
CorelDraw 74, 105
CPS-1 *see* Capcom Play
    System One
CRTs *see* cathode ray tubes

D

D-Paint 82
developers 78–131
    amateur 159
    Cabybara 98–103
    character development 98–9,
    104–7
    file formats 90–1, 102

Appendix

forced-perspective 98
Halfer, Riva Jan 110–11
McWhertor, Michael 114–15
memory usage 80, 82, 90–2,
    96, 116
porting 94, 118–31
production cycle 94–5
resolution 92–3, 119, 122
sprite creation 100–3, 109–15
sprite formats 88–9
Takayoshi, Sato 112–13
development tools 30–1, 56–9, 82,
    88–9
digital photographs 43
digitized graphics 10
dithering 82
Donkey Kong 154, 156, 162, 164–7
Donkey Kong Junior 16–17, 114, 164
Dragon Quest 134–9
Dungeons and Dragons 134

E

eBoy 68–71
Exodus 133, 136

F

fighting games 146–53
file formats 90–1, 102
Film Strip 88
Final Fantasy 138–44
Flash 72–4, 108
forced-perspective 98
frame rates 9
frontlighting 51

G

Game & Watch 15, 16–19
GameBoy
    competitors 24, 26, 34
    developers 119
    history 15, 20–1
    platform games 162, 165, 173–4

rendering 52
role-playing games 182
GameBoy Advance
    competitors 30, 36, 49
    history 19
    platform games 158–9, 170–1
    porting games from SNES 124–5
    role-playing games 182
GameGear 22, 52, 126–7, 158, 169
Gameloft 84–7
gameplay 10, 80, 150–1
Genesis see Sega MegaDrive
Ghouls 'n' Ghosts 125, 128–31,
    176–9
Glu Mobile 79, 82–4, 86
Grafx2 59
granularity 57
Graphics Gale 59

H

Halfer, Riva Jan 110–11
hardware 42–77
    Army of Trolls 64–7
    color displays 42–3, 48–51
    development tools 56–9
    eBoy 68–71
    Hildenbrand, Chris 72–7
    motion blurring 52–3
    Nieborg, Henk 60–3
    resizing pixels 54–5
    see also cellphones;
        portable devices
head-to-head play 146
Heap space 90, 92
Hildenbrand, Chris 72–7, 104–7
Human Balance 59

I

i-mode 130, 176
ImageReady 64
independent developers 87

J

Jadestone 80–1, 86
JIT see Just in Time
joysticks 146, 148–9
Jumbled-Up Piece of Crap
    (JUPOC) 88–9
Just in Time (JIT) 90

K

Karate Champ 146–7

L

LaserDiscs 10
LCDs see liquid crystal displays
leading-edge erosion 53
light emitting diodes (LEDs) 15
light sources 99
liquid crystal displays (LCDs) 50–1
Lucken, Gary 64–7
Lynx 28–9, 34–5

M

McPherson, Ian 116–17
Macromedia Flash 72–4, 108
McWhertor, Michael 114–15
Marios 64 114–15
Master System 126–7, 155, 158,
    168–9
MegaDrive see Sega MegaDrive
Megaman 54–5
memory usage 80, 82, 90–2, 96, 116
Microvision 14
Misadventures of Flink, The 60
Miyamoto, Shigeru 164
Mortal Kombat 158
motion blurring 52–3, 126

N

N-Gage 122
NeoGeo MVS 152–3
NeoGeo Pocket 24–7, 35

Index

NES *see* Nintendo Entertainment
 System
Nieborg, Henk 60–3
Nintendo 64, 158, 163
Nintendo DS 36–7, 159, 182
Nintendo Entertainment System (NES)
 developers 119, 128
 history 20–1, 34
 platform games 155, 161, 173
 role-playing games 128, 136, 176,
  180, 182
 see also Super Nintendo
  Entertainment System
NTT DoCoMo 130, 176

O

off-axis viewing 50
organic light emitting diodes
 (OLEDs) 50
overhead view 144

P

PacMan 26, 56–7, 84, 113, 154
parallax scrolling 155
PC Engine 22, 116–17, 155, 182
personal digital assistants (PDAs) 6
perspective 98
Pettersson, Henrik 80, 82, 86
Photoshop 64, 66, 82, 99–101
pixels 8–13
 artists 60–77
 cellphones 45, 46, 48, 78, 82–4,
  86–7
 character development 160–83
 development tools 56–9
 motion blurring 52–3, 126
 pixel toys 40–1
 portable devices 11, 32, 36, 38
 resizing 54–5
 resolution 84, 86, 92–3
 sprite creation 100–3, 109–15
 vectors 12–13

platform games 154–9
PlayStation 6, 8, 157–8, 182
PlayStation Portable (PSP) 32, 35,
 38–9, 53
PNG format 90, 102
polygons
 cellphones 78
 current generation devices 32
 developers 56, 78, 119
 hardware 56
 history 9, 10–11
 platform games 156, 166
 sprite formation 61, 76
Pong 42
portable devices 6, 14–39
 Game & Watch 15, 16–19
 Lynx 28–9, 34–5
 NeoGeo Pocket 24–7, 35
 Nieborg, Henk 62
 Nintendo DS 36–7, 159, 182
 pixel toys 40–1
 pixels 11, 32, 36, 38
 PlayStation Portable 32, 35,
  38–9, 53
 Turbo Express 22–3, 35
 WonderSwan 18, 30–1, 179
 see also cellphones; GameBoy;
  GameBoy Advance
porting 94, 118–31, 168–9
pre-rendering 13, 56, 144, 156,
 158, 166
Pro Motion 82
production cycle 94–5
PSP *see* PlayStation Portable

R

raster monitors 12–13
Rayman 157
rendering
 developers 113
 pixel art 56
 platform games 156, 158, 166

role-playing games 144
vector graphics 13
resizing 118–21, 157–8
 aspect ratios 122–3, 125, 178
 Cabybara 99
 cellphones 94, 96
 pixels 54–5
 porting 123
resolution 92–3, 119, 122
retro gaming 11, 45, 78, 84
role-playing games (RPGs) 18, 131,
 132–45

S

saturation 48, 50, 51
Saturn (Sega) 157–8
scaling see resizing
scrolling levels 155
Sega Master System 126–7, 155,
 158, 168–9
Sega MegaDrive
 competitors 35
 developers 120, 127, 129
 Nieborg, Henk 60
 platform games 157, 158, 170
 porting to GameGear 127
 role-playing games 129, 178
Sega Saturn 157–8
sidescrolling view 144
Snake 48
SNES see Super Nintendo
 Entertainment System
soft scaling 54–5
Sonic the Hedgehog 120, 127, 155,
 156, 168–71
Sony PlayStation 6, 8, 157–8, 182
Sony PlayStation Portable (PSP) 32,
 35, 38–9, 53
Space Invaders 42, 84
sprite formats 88–9
stippling 170
Street Fighter 146–52

Super Ghouls 'n' Ghosts 130–1
Super Mario Brothers
 developers 84, 113, 114–15, 119
 Game & Watch 16
 GameBoy 20
 motion blurring 52
 pixel toys 40–1
 porting 119
 retro gaming 84
 scrolling levels 155
 sprite history 160–3
 transparency 156
Super Nintendo Entertainment
 System (SNES)
 developers 130
 platform games 156, 158, 162,
  166, 173–5
 porting to GameBoy Advance 124–5
 role-playing games 130, 139
 see also Nintendo Entertainment
  System
SuperGrafx-16 129

T

Takayoshi, Sato 112–13
Tandy TRS-80 132
Tetris 21, 34, 53
transparency 156, 157–8, 170
TurboExpress 22–3, 35
TurboGrafx-16 22–3, 35, 58, 155,
 158, 172

U

UIEvolution 80
Ultima III: Exodus 133, 136

V

vacuum fluorescent displays 15
Van Der Vlag, Randy 84, 86–7
vectors 12–13, 93, 109
Vella, Nathan 98–103
Virtual Boy 15, 18

W

Wonder Witch 30–1
WonderSwan 18, 30–1, 179

Y

Yie-Ar Kung Fu 148–9
Yokoi, Gunpei 17–18, 30

Z

Zirkle, Paul 87–97

Thanks

LINKS

http://pixeljoint.com/
Perhaps the best public gallery for amateur and professional pixel artists.

www.natomic.com/hosted/marks/mpat/
An excellent set of tutorials for the budding pixel artist.

www.zoggles.co.uk/asp/tutorials.asp?tut=1
More excellent tutorials

http://rhysd.syntesis.org/tutorial/
Excellent isometric pixel tutorial

www.spriteattack.com/
Chris Hildenbrand's site

http://segasaturn.de
Excellent vector art

http://nfg.2y.net/
The author's site

Thanks

The list of people who deserve thanks is a long one. Nothing I do is done in a vacuum, there are giants on whose shoulders I stand. My parents. They made me, they made me who I am. Thanks guys, I love you. Zumi, baby, you're the best. Aishiteru. My friends, past and present, for encouraging me and for hiding their laughter.

Big thanks to everyone who contributed to the making of this book:

Brandon Sheffield and Matt Warner
Chris Hildenbrand
Glu Mobile
Blue Label Games
Capybara Games
UIE
I-Play
Game Loft
Jadestone

Thanks to all the people who made the games we love to play.

Thanks also to: Dean Scott and everyone else who gave me a voice.
What were you thinking?

The guys who followed me into #NFG. Suckers. Spekkio, Albx, merrykan, bloodf, C-Storm, corpsicle, elend, jef, manu, MrNES (0uch!), nolanXL, Outrider, Stt, Twyst, Tychom, V3rtigo, Wehr, ameneko, DJH, Funky, Sixtoe, Steempy.

Appendix